Muhammad Shafiq
Fouzia Yaqub
Isma Younes

# Spatiotemporal Variations in Urban Air Quality of Lahore, Pakistan

Anchor Academic
Publishing

Shafiq, Muhammad, Yaqub, Fouzia, Younes, Isma: Spatiotemporal Variations in Urban
Air Quality of Lahore, Pakistan, Hamburg, Anchor Academic Publishing 2016

Buch-ISBN: 978-3-95489-492-5
PDF-eBook-ISBN: 978-3-95489-992-0
Druck/Herstellung: Anchor Academic Publishing, Hamburg, 2016

**Bibliografische Information der Deutschen Nationalbibliothek:**
Die Deutsche Nationalbibliothek verzeichnet diese Publikation in der Deutschen
Nationalbibliografie; detaillierte bibliografische Daten sind im Internet über
http://dnb.d-nb.de abrufbar.

**Bibliographical Information of the German National Library:**
The German National Library lists this publication in the German National Bibliography.
Detailed bibliographic data can be found at: http://dnb.d-nb.de

All rights reserved. This publication may not be reproduced, stored in a retrieval system
or transmitted, in any form or by any means, electronic, mechanical, photocopying,
recording or otherwise, without the prior permission of the publishers.

Das Werk einschließlich aller seiner Teile ist urheberrechtlich geschützt. Jede Verwertung
außerhalb der Grenzen des Urheberrechtsgesetzes ist ohne Zustimmung des Verlages
unzulässig und strafbar. Dies gilt insbesondere für Vervielfältigungen, Übersetzungen,
Mikroverfilmungen und die Einspeicherung und Bearbeitung in elektronischen Systemen.

Die Wiedergabe von Gebrauchsnamen, Handelsnamen, Warenbezeichnungen usw. in
diesem Werk berechtigt auch ohne besondere Kennzeichnung nicht zu der Annahme,
dass solche Namen im Sinne der Warenzeichen- und Markenschutz-Gesetzgebung als frei
zu betrachten wären und daher von jedermann benutzt werden dürften.

Die Informationen in diesem Werk wurden mit Sorgfalt erarbeitet. Dennoch können
Fehler nicht vollständig ausgeschlossen werden und die Diplomica Verlag GmbH, die
Autoren oder Übersetzer übernehmen keine juristische Verantwortung oder irgendeine
Haftung für evtl. verbliebene fehlerhafte Angaben und deren Folgen.

Alle Rechte vorbehalten

© Anchor Academic Publishing, Imprint der Diplomica Verlag GmbH
Hermannstal 119k, 22119 Hamburg
http://www.diplomica-verlag.de, Hamburg 2016
Printed in Germany

# Preface

This study reports the air pollution vulnerability mapping of Lahore city and provides information about the vulnerability levels of air pollutants at various places of Lahore. The air quality monitoring activities have been carried out at five places of Lahore and each location is monitored for six air pollutants which are carbon monoxide, sulphur dioxide, nitrogen dioxide, hydrogen sulphide, ammonia and chlorine by a reagent absorption method. The sources and concentrations have been discussed. The data obtained from sampling at various commercial and residential sites is compared against the past year data and the comparison graphs are plotted to show the trend of these air pollutant. The spatial patterns of monthly mean precipitation and monthly mean temperature have been studied. A public survey has been conducted that shows the public awareness in regard to basic information regarding air pollution. In the end some conclusions and recommendations are given for the purpose to maintain a healthy environment for our citizens, support the strategies, to maintain energy independence and promote an effective role in formulating the environmental policies.

# Contents:

Preface ............................................................................................................. i
Contents: ......................................................................................................... iii
Chapter 1: Introduction ................................................................................. 1
   1.1    Study area ............................................................................................ 2
   1.2    Climatological Conditions ................................................................. 4
        1.2.1 Spatial Patterns of Precipitation in the Punjab Province ............ 4
        1.2.2 Spatial Patterns of Temperature in the Punjab Province ......... 19
        1.2.3 Humidity .................................................................................. 33
        1.2.4 Wind Direction ........................................................................ 33
   1.3    Population Growth of Lahore ........................................................... 33
   1.4    Existing Land Use ............................................................................ 34
   1.5    Air pollution definition .................................................................... 36
   1.6    Common air pollutants .................................................................... 37
   1.7    Major sources of air pollution ......................................................... 37
   1.8    Concerns to air pollution ................................................................. 38
        1.8.1   Should we not worry about air pollution ............................. 38
   1.9    An insight into Int'l (Asian) air pollution levels ............................. 38
        1.9.1   Air pollution in China ........................................................... 39
        1.9.2   Air pollution in India ............................................................ 40
        1.9.3   Air pollution in Indonesia ..................................................... 40
        1.9.4   Air pollution in Hong Kong .................................................. 40
   1.10   The role of International Cooperation ............................................ 41

Chapter 2: Classification, sources and effects ............................................ 43
   2.1    Type A: Classification ..................................................................... 43
        2.1.1   Natural sources (primary and secondary pollutants) ............ 43
        2.1.2   Manmade sources .................................................................. 44
   2.2    Type B: Classification ..................................................................... 44
        2.2.1   Point sources .......................................................................... 44
        2.2.2   Line sources .......................................................................... 45
        2.2.3   Area / volume sources .......................................................... 45
   2.3    Sulphur containing compounds ....................................................... 45
        2.3.1   Sulphur dioxide ..................................................................... 45
        2.3.2   Sulphur trioxide ..................................................................... 46
        2.3.3   Hydrogen sulphide ................................................................ 46
   2.4    Nitrogen containing compounds ..................................................... 47
        2.4.1   Nitrous oxide ........................................................................ 47
        2.4.2   Nitric oxide ........................................................................... 47

    2.4.3    Nitrogen dioxide ................................................................................47
2.5    Carbon containing compounds.................................................................48
    2.5.1    Carbon monoxide...............................................................................48
    2.5.2    Carbon dioxide...................................................................................49
2.6    Residential (indoor) air pollution sources ..............................................50
2.7    Effects of air pollution..............................................................................51
    2.7.1    Effects of air pollution on atmospheric properties .....................51
    2.7.2    Effects of air pollution on human health ......................................52
    2.7.3    Airborne diseases...............................................................................54
    2.7.4    Air pollution effects on the materials ............................................55
2.8    Effects on plants ........................................................................................56
2.9    Types of plants in Lahore .........................................................................57
    2.9.1    Roadside Plants..................................................................................57
    2.9.2    Ediphytes and Chasmophytes.........................................................58
2.10    Sensitivity of the plants to air pollution ................................................59
    2.10.1    Damage to Leaf Structure...............................................................59
    2.10.2    Delayed Flowering..........................................................................59
    2.10.3    Root Damage ...................................................................................60
2.11 Residence time of air pollutants ...................................................................60

Chapter 3: Methods .................................................................................................63
3.1    The air pollution monitoring unit ...........................................................63
    3.1.1    Equipment requirement ....................................................................63
    3.1.2    Selection of a sample site ................................................................64
    3.1.3    Setting the sampling point ...............................................................65
3.2    Objectives and scope of the kit fabrication program ...........................65
3.3    The air pollution monitoring unit ...........................................................65
    3.3.1    Monitoring parameters.....................................................................67
    3.3.2    The air pump.....................................................................................67
    3.1.3    Air pollution monitoring setup .......................................................68
    3.3.4    Plastic and glassware ........................................................................68
    3.3.5    Composition of reagents for detecting gases................................68
3.4    Procedure for air pollution analysis: .......................................................70
    3.4.1    Carbon monoxide...............................................................................70
    3.4.2    Nitrogen dioxide ................................................................................71
    3.4.3    Sulphur dioxide..................................................................................72
    3.4.4    Hydrogen sulphide.............................................................................73
    3.4.5    Ammonia ............................................................................................74
    3.4.6    Chlorine ..............................................................................................75
3.5    Precautions .................................................................................................76
    3.5.1    The air pump:.....................................................................................76
    3.5.2    The monitoring unit:.........................................................................76
3.6    Troubleshooting.........................................................................................77
3.7    Visual comparators....................................................................................77

## Chapter 4: Air pollution sampling and analysis ............ 81

- 4.1 Importance of monitoring air pollution ............ 81
- 4.2 Importance of data acquisition and analysis ............ 82
- 4.3 Exposure level study procedure ............ 84
  - 4.3.1 Human exposure studies ............ 84
  - 4.3.2 Laboratory research / animal studies ............ 84
  - 4.3.3 Epidemiological studies ............ 85
- 4.4 WHO concentration thresholds ............ 85
- 4.5 Atmospheric concentration units ............ 86
- 4.6 Air pollution at Badami Bagh ............ 87
  - 4.6.1 Map of the sampling site ............ 87
  - 4.6.2 Data acquisition and plotting ............ 88
  - 4.6.3 Interpretation and trend analysis ............ 91
- 4.7 Air pollution at Ichhra Market ............ 92
  - 4.7.1 Map of the sampling site ............ 92
  - 4.7.2 Data acquisition and plotting ............ 93
  - 4.7.3 Interpretation and trend analysis ............ 96
- 4.8 Air pollution at Main Market Gulberg ............ 97
  - 4.8.1 Map of the sampling site ............ 97
  - 4.8.2 Data acquisition and plotting ............ 98
  - 4.8.3 Interpretation and trend analysis ............ 101
- 4.9 Air pollution at Railway Station ............ 102
  - 4.9.1 Map of the sampling site ............ 102
  - 4.9.2 Data acquisition and plotting ............ 103
  - 4.9.3 Interpretation and trend analysis ............ 106
- 4.10 Air pollution at Shadman Market ............ 107
  - 4.10.1 Map of the sampling site ............ 107
  - 4.10.2 Data acquisition and plotting ............ 108
  - 4.10.3 Interpretation and trend analysis ............ 111
- 4.11 Peak concentrations (in ppm) ............ 112
  - 4.11.1 Interpretation and trend analysis ............ 115
- 4.12 Mean concentrations (in ppm) ............ 116
  - 4.12.1 Mean of past pollutant concentrations (in ppm) ............ 117
  - 4.12.2 Interpretation and trend analysis ............ 121
  - 4.12.3 Interpretation and inter-comparison ............ 122
- 4.13 Lahore city averages (in ppm) ............ 123
  - 4.13.1 Interpretation and trend analysis ............ 124

## Chapter 5: Air pollution control ............ 125

- 5.1 Gaseous pollution control methods ............ 125
  - 5.1.1 Absorption and adsorption towers ............ 125
  - 5.1.2 Combustion ............ 126
- 5.2 Particulate pollution control methods ............ 126
  - 5.2.1 Settling chamber ............ 127

|       | 5.2.2  | Cyclone separator ..................................................................... 128 |
|       | 5.2.3  | Wet collectors (Scrubbers) ....................................................... 129 |
|       | 5.2.4  | Bag filters.................................................................................. 132 |
|       | 5.2.5  | Electrostatic precipitators ........................................................ 132 |

Chapter 6: Public survey, conclusions and recommendations ........................... 135
    6.1    Public survey ................................................................................... 135
    6.2    Conclusions and recommendations ................................................. 143

Bibliography ........................................................................................................ 151

# Chapter 1:
# Introduction

Air pollution is an increasingly important environmental problem in the world. Emissions of sulphur dioxide, nitrogen oxides and ammonia have been rising steadily over the past few decades.

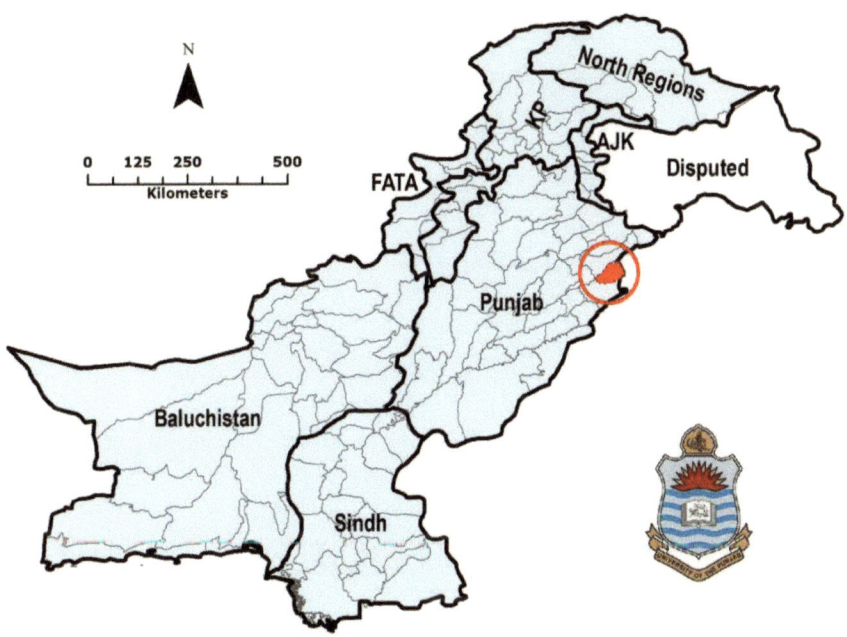

Figure 1.1: Location of the study area (Lahore City) is shown in red circle

Projections indicate that potentially large increases in emissions may occur during the next twenty to fifty years if current development patterns persist. If these occur, the impacts that have been experienced in Europe over this century will become increasingly prominent in large parts of Asia over the next century. As an initiative to facilitate the development of action plans, strategies and policies for pollution prevention and control, many countries are funding many programs on atmospheric and environmental issues in developing countries.

## 1.1 Study area

Lahore is a typical inland city of Pakistan. As shown in Figure 1.1, this land locked city is situated about 1100 kilometres away from the Arabian Sea.

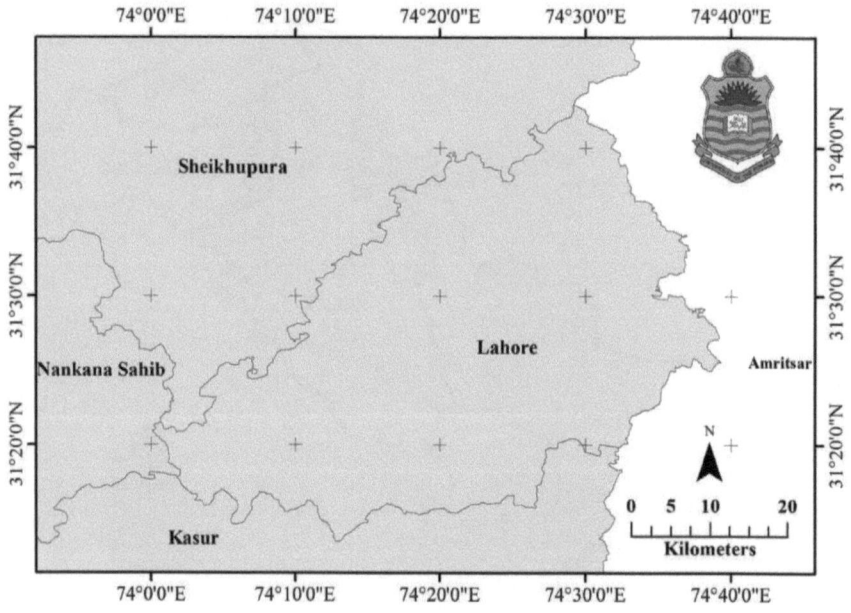

Figure 1.2: Geographical Location of Lahore City

The geographical location of Lahore is from 31° 13′ 28″ N to 31° 43′ 02″ N in latitude and 74° 00′ 00″ E to 74° 39′ 05″ N in longitude. Situated along the South bank of River Ravi, the city is bounded by Sheikhupura district in the North and Kasur District in the South. The east of Lahore city is the International Boundary Line separating Pakistani Punjab from Indian Punjab. The adjoining city on the Indian side is Amritsar (Ajnala, Tarn Taran and Patti) as shown in Figure 1.2. With a total area of 1,772 square kilometres, the Lahore city is the provincial capital of Punjab and its large area comprises of urban settlements as shown in Figure 1.3.

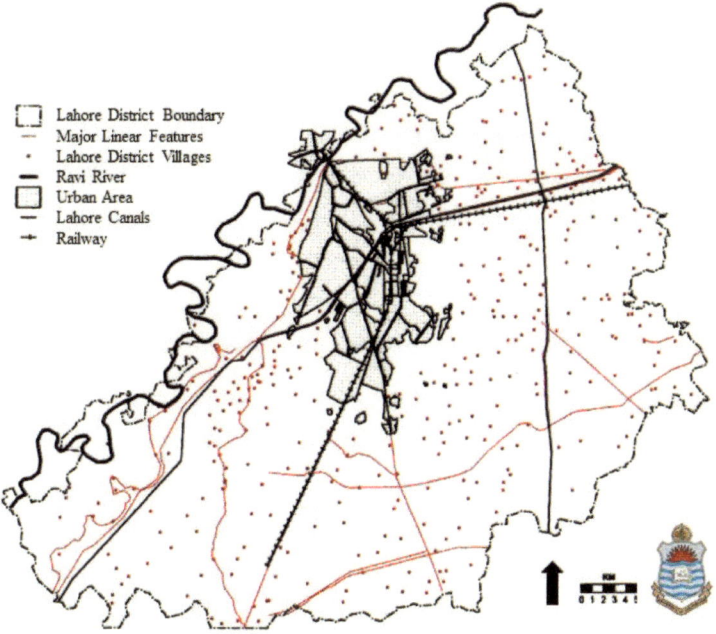

Figure 1.3: Urban area of Lahore district (Isma, 2000)

Lahore occupies a fertile alluvial plain formed by the deposits of the river Ravi and Sutlej and their tributaries. The river flows along the north-western boundary of the city, and its seasonal floods have inhabited the city's growth towards the North and Northwest.

## 1.2 Climatological Conditions

Climatically, the city of Lahore experiences three seasons, cold, hot, and rainy. Winter season extends from mid-November to the mid-February. In this season the cold sunny days are alternated with cloudy rainy days. Rainy days are usually followed by clear days with frost at night in the countryside. Early morning and evening mist is also common.

The winter rain usually stops by the end of February but sometimes continue through March. The monsoon burst over Lahore sometimes in the mid of the June or the last week of the June. It usually results in the heavy downpour and the showers continue with the short intervals till September.

### 1.2.1 Spatial Patterns of Precipitation in the Punjab Province

Most of the rainfall in Lahore occurs in the monsoon months. The monsoons are at their peak in July. In the rainy season, it is pleasant, but during the day following rains, the still air, high temperature, and high humidity create a condition of extreme discomfort. On the average there is no month without rainfall in Lahore. The driest months are October and November, which together receive on an average 0.34 inches of the rainfall. From December to March is the season of winter rains, which the entire four months period receive 3.72 inches of rainfall an amount although meagre, but of great importance for its value to agriculture and the weather. April and May receive rainfall either through western disturbances or locally generated thunderstorms. May to October account for only about 25% of the annual rainfall. The remaining 75% takes place in the monsoon season.

The monsoons are at their peak during July and August. On the average July is the month of the maximum and November is of minimum rainfall. The most intensive recorded cloudburst in Lahore occurred on September 24, 1954 which about 9 inches rainfall in 24 hours. Such cloudbursts seriously affect the daily life of the city by flooding the low-lying areas due to inadequate drainage system. The heaviest rainfall on Lahore occurred in the 1882 when recorded 37.43 inches of rainfall was recorded. The driest year was 1899 when only 6.21 inches of rainfall.

Here the data from WorldClim database has been used to identify the estimates of the spatial patterns of precipitation in the Punjab province. As the WorldClim data grids are of 1 kilometres resolution, a small city may contain only a few pixels. In order to better visualize the patterns it is important to consider / map a larger area. Therefore, the precipitation maps have been prepared on the province level scale.

The WorldClim database is based on the data interpolated over the larger areas (near global) from obtained from the multiple weather stations (WMO) during the period 1960-90. The resampled SRTM digital elevation model data at a resolution of 1 kilometre has been used for spline interpolating the point data over the larger spatial surfaces.

Traditionally the units for the measurement of rainfall are mm (millimetre) which is the amount of rain per square meter in one hour. The amount of 1mm of rainfall means one litre of water in a square meter area. The unit used for the precipitation values here is in mm equivalent of rainfall.

The maps of rainfall in the Punjab province indicate heavy rainfalls in the northern part which is adjacent to the mountainous northern areas of Pakistan comprising of State of Azad Jammu and Kashmir (AJK), North West Frontier Province (NWFP), and Shumali Ilaqajat. The southern parts of Punjab receive very little rainfall.

The rainfall values presented here comprise of modelled data and may contain inaccuracies and/ or over-estimations. The values of monthly cumulative precipitation averaged for 1060-90 period (lowest and highest values of rainfall in the Punjab province) are summarized in the following table.

| Month | Monthly Rainfall (lowest value in Punjab) in mm | Monthly Rainfall (highest value in Punjab) in mm |
|---|---|---|
| January | 2 | 109 |
| February | 2 | 112 |
| March | 3 | 137 |
| April | 1 | 112 |
| May | 1 | 83 |
| June | 4 | 107 |
| July | 23 | 255 |
| August | 24 | 255 |
| September | 8 | 179 |
| October | 0 | 55 |
| November | 0 | 29 |
| December | 2 | 47 |

Table 1.1: Seasonal variation of precipitation in the Punjab province

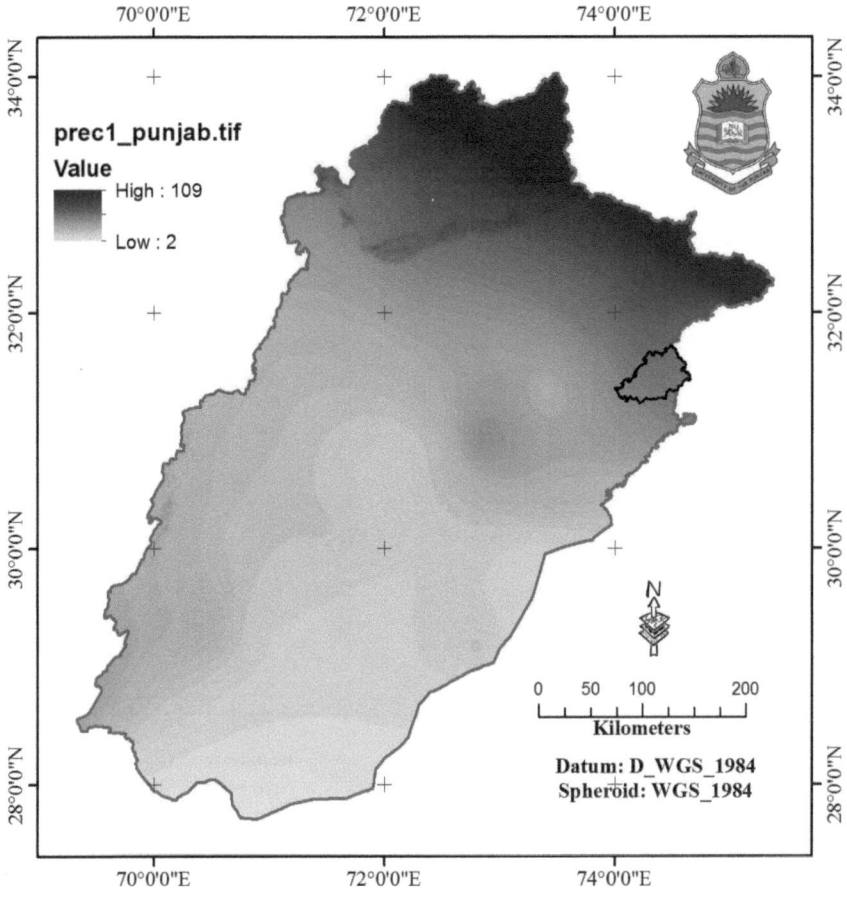

Figure 1.4: Spatial distribution of mean monthly precipitation (in mm) for the month of January for Punjab province

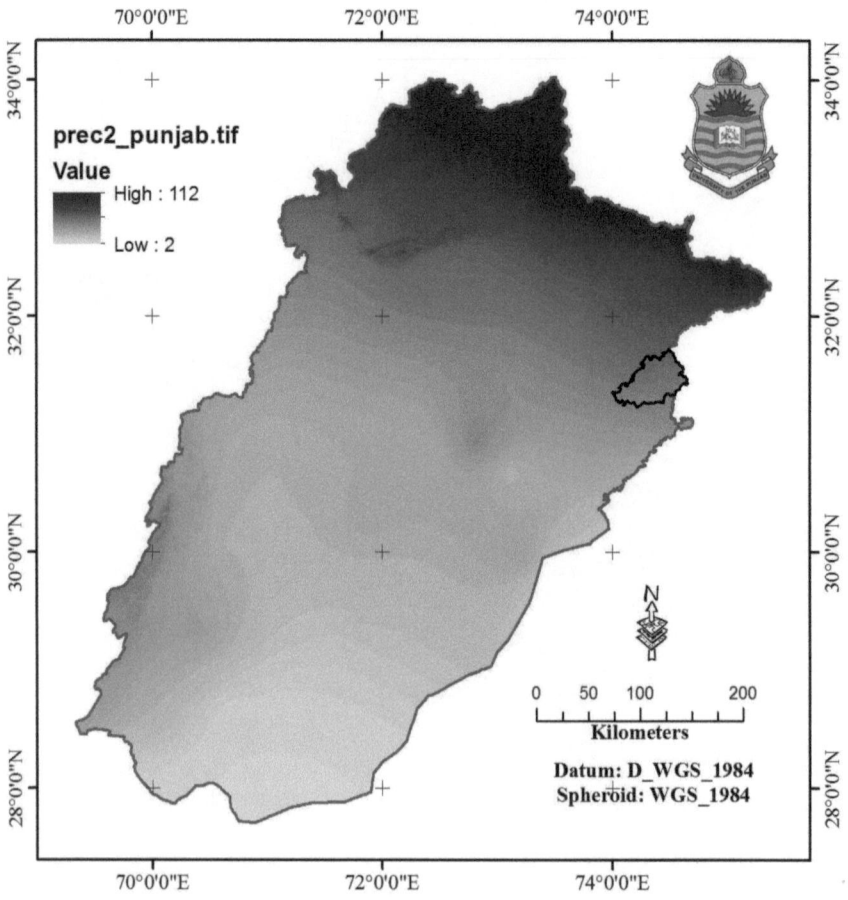

Figure 1.5: Spatial distribution of mean monthly precipitation (in mm) for the month of February for Punjab province

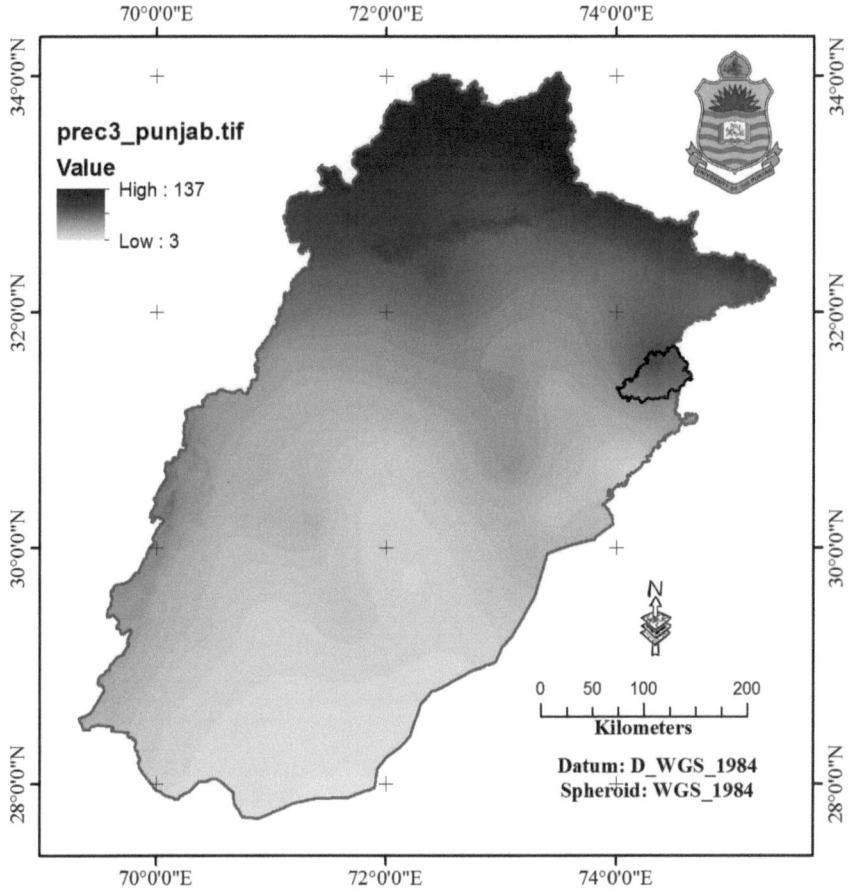

Figure 1.6: Spatial distribution of mean monthly precipitation (in mm) for the month of March for Punjab province

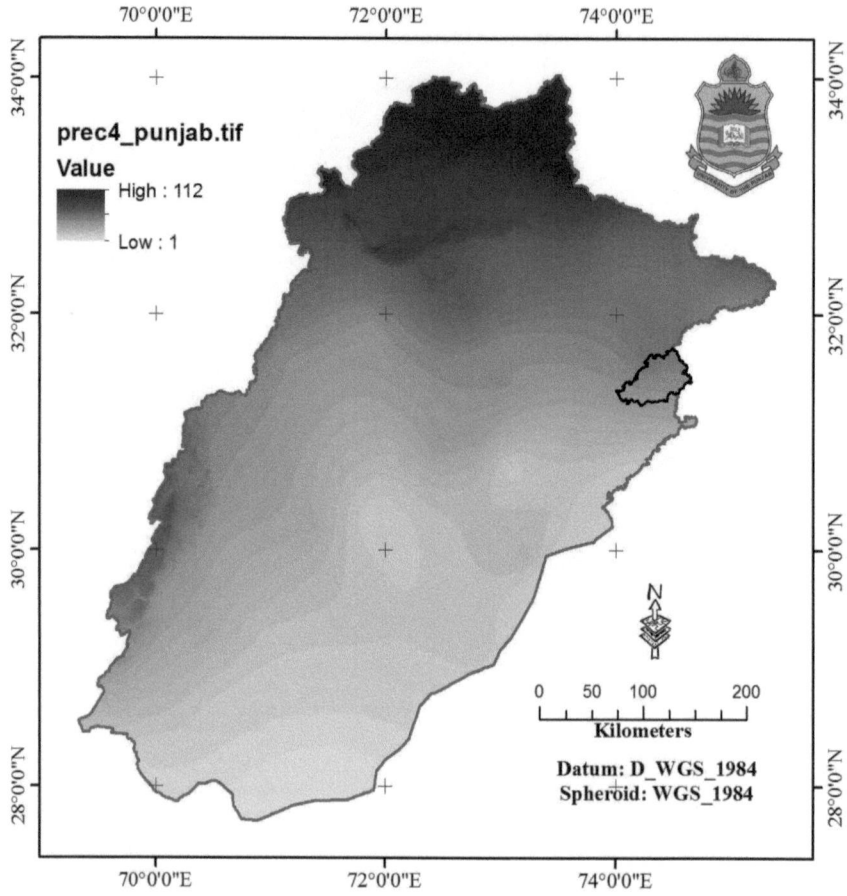

Figure 1.7: Spatial distribution of mean monthly precipitation (in mm) for the month of April for Punjab province

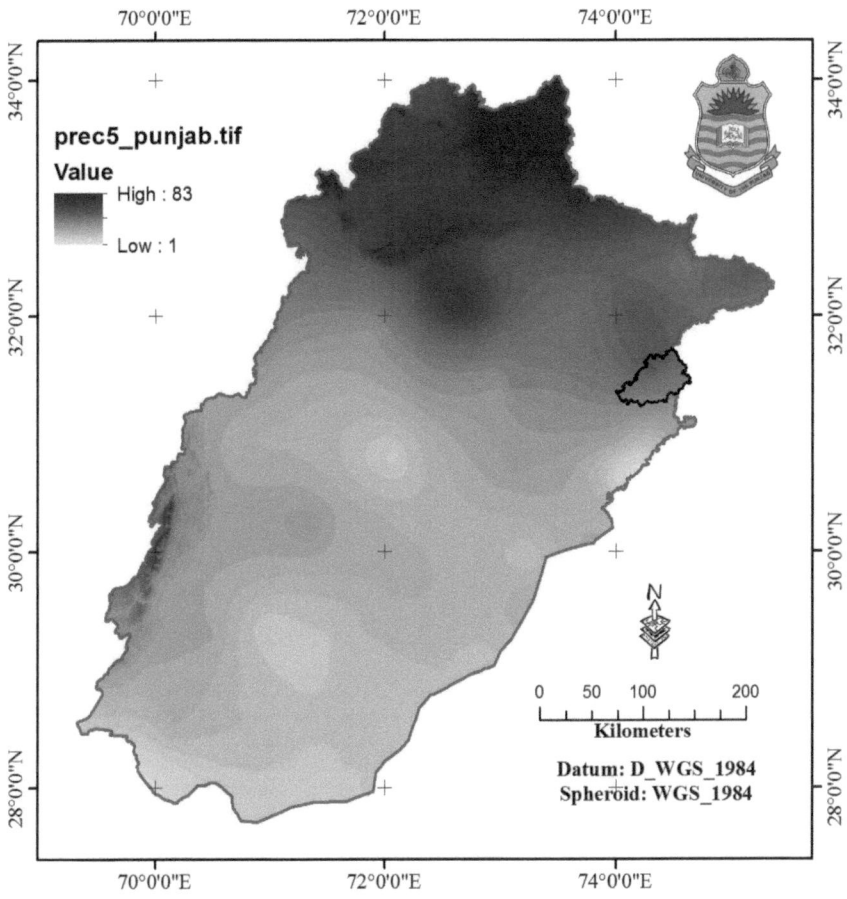

Figure 1.8: Spatial distribution of mean monthly precipitation (in mm) for the month of May for Punjab province

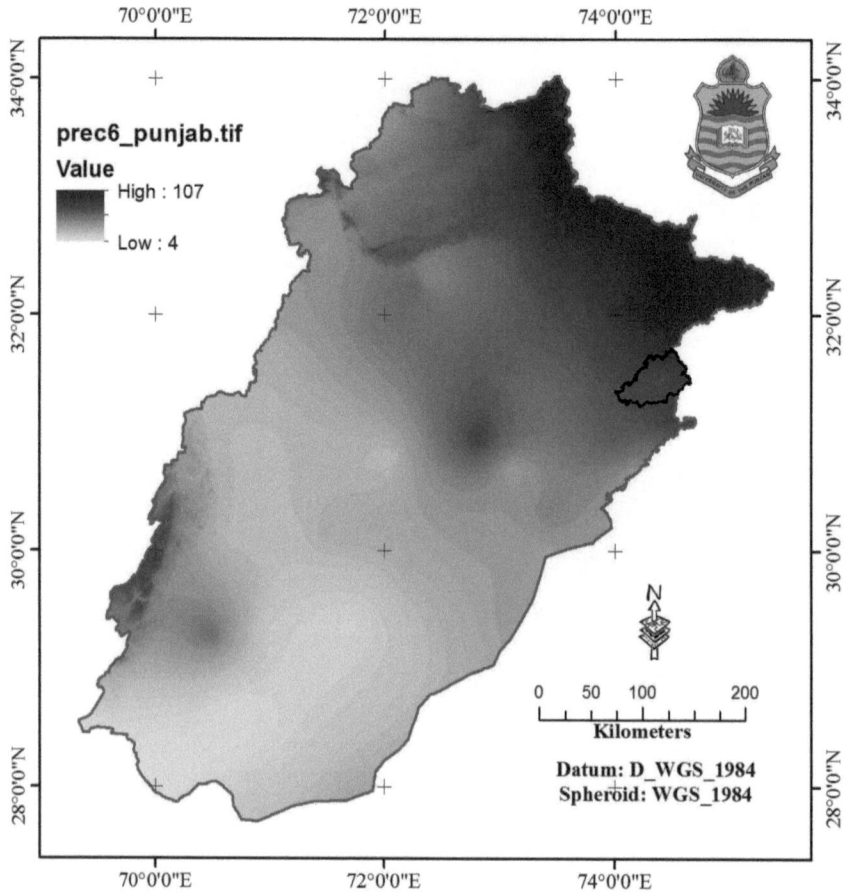

Figure 1.9: Spatial distribution of mean monthly precipitation (in mm) for the month of June for Punjab province

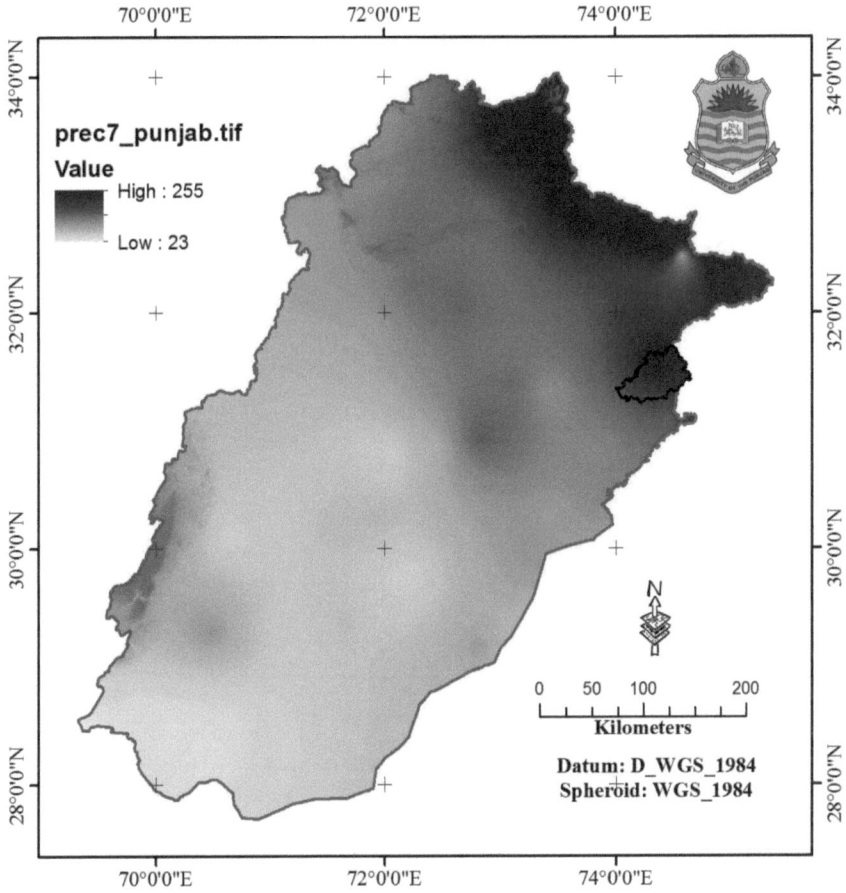

Figure 1.10: Spatial distribution of mean monthly precipitation (in mm) for the month of July for Punjab province

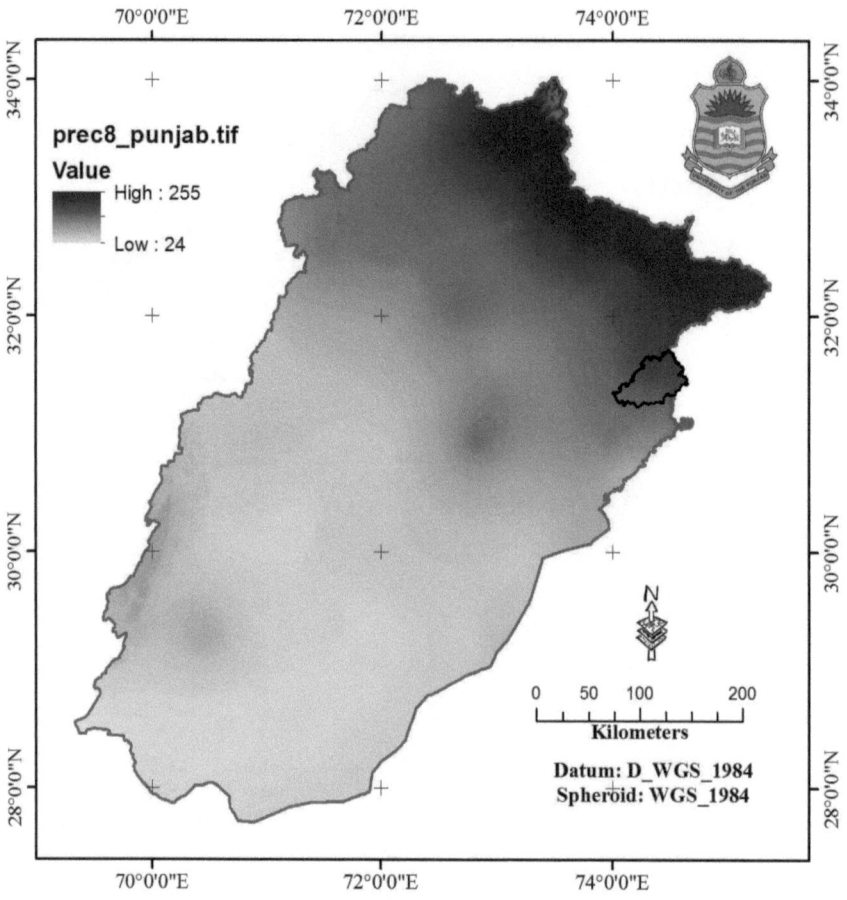

Figure 1.11: Spatial distribution of mean monthly precipitation (in mm) for the month of August for Punjab province

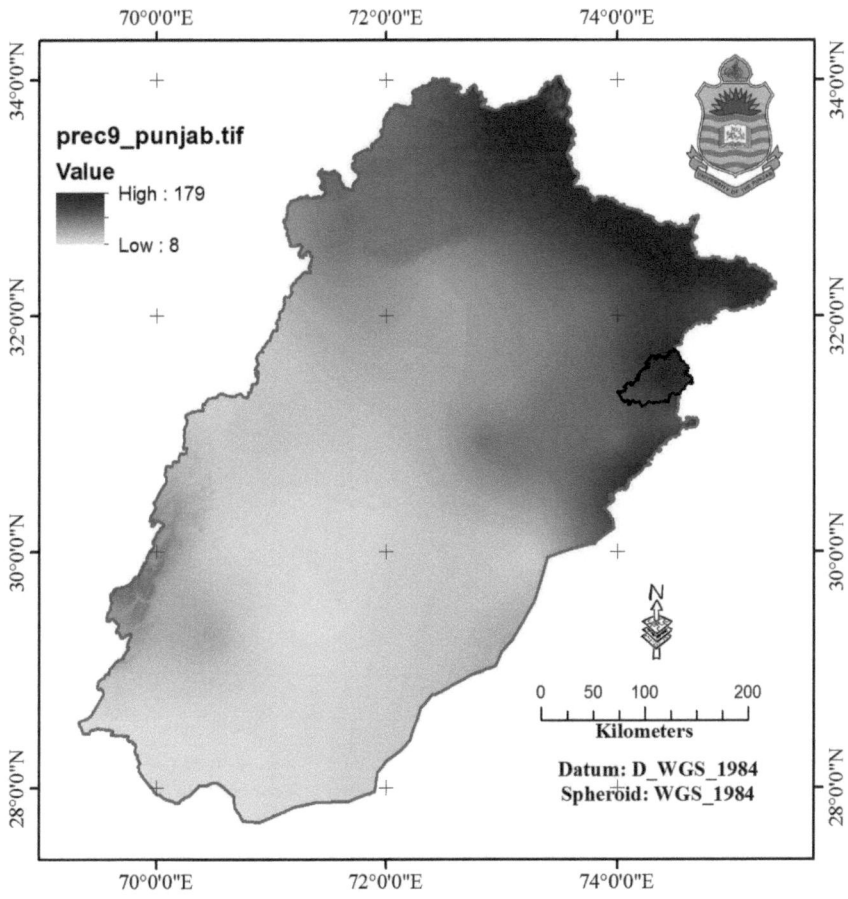

Figure 1.12: Spatial distribution of mean monthly precipitation (in mm) for the month of September for Punjab province

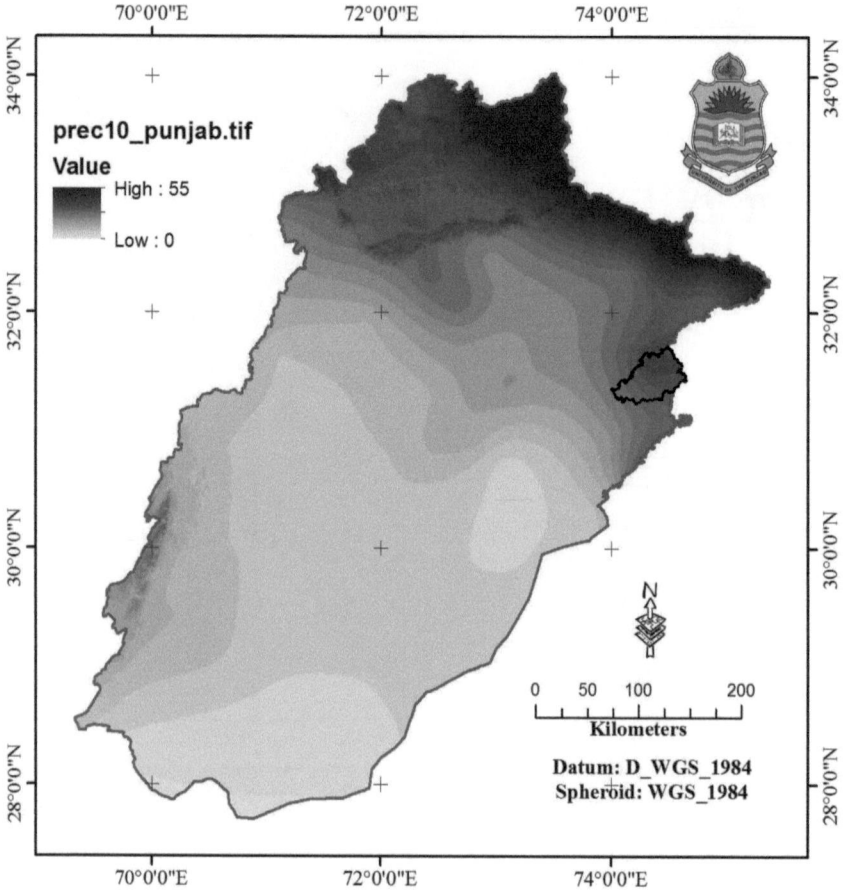

Figure 1.13: Spatial distribution of mean monthly precipitation (in mm) for the month of October for Punjab province

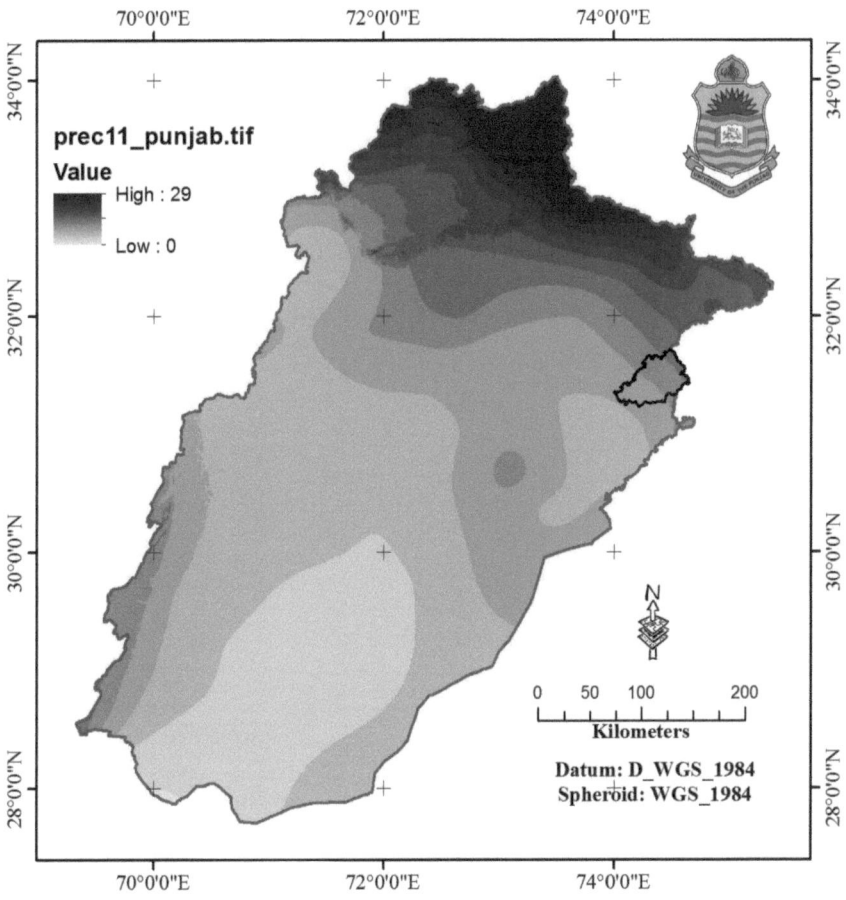

Figure 1.14: Spatial distribution of mean monthly precipitation (in mm) for the month of November for Punjab province

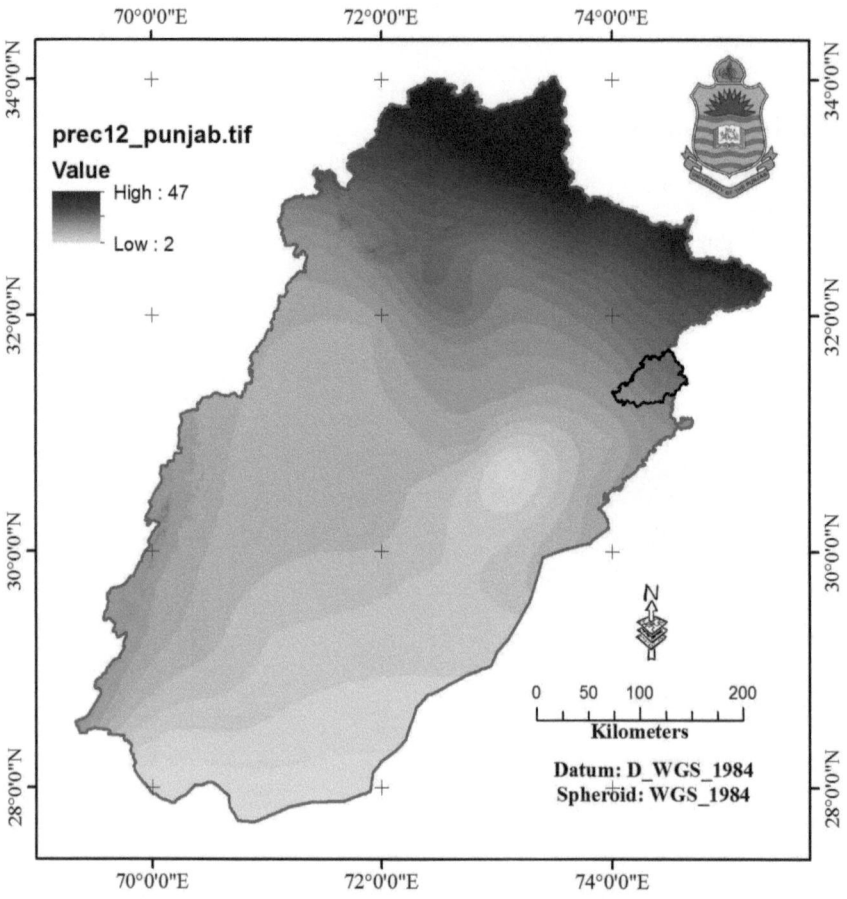

Figure 1.15: Spatial distribution of mean monthly precipitation (in mm) for the month of December for Punjab province

## 1.2.2 Spatial Patterns of Temperature in the Punjab Province

Temperatures remain comfortable in this zone. The average of minimum temperature of January is 66 °F. On very rare occasions the night temperature falls below freezing point. The lowest temperature ever recorded is 40 °F May and June constitute the hot season of which June is the hottest month. The day temperature rises to scorching level, and due to hot breeze, is instantly uncomfortable. The average minimum temperature in June is 80 °F, though the highest of the season may touch 116 °F mark. In July, during the day following rains, high temperature is experienced. January is the coldest month.

Here the monthly mean temperature data from WorldClim database has been used to identify the estimates of the spatial patterns of temperature in the Punjab province. As the WorldClim data grids are of 1 kilometres resolution, a small city may contain only a few pixels. In order to better visualize the patterns it is important to consider / map a larger area. Therefore, the temperature maps have been prepared on the province level scale.

The WorldClim database is based on the temperature data interpolated over the larger areas (near global) from obtained from the multiple weather stations (WMO) during the period 1960-90. The resampled SRTM digital elevation model data at a resolution of 1 kilometre has been used for spline interpolating the point data over the larger spatial surfaces.

Traditionally the units for the measurement of temperature are degrees Celsius of which the range is from 0 to 100. The monthly mean temperature data comes in degree Celsius times 10 ($^0C \times 10$). This means a value of 140 represent 14 $^0C$.

The maps of mean monthly temperature averaged over the period of 1960-90 for the Punjab province indicate higher temperatures over the southern part of Punjab province. The northern part of Punjab province comprises of mountainous terrain and is characterized by lower temperatures due to elevation effect and temperature gradients. The temperature values presented here comprise of modelled data and may contain inaccuracies and/ or over/under-estimations.

The mean monthly values of minimum and maximum values of averaged temperature in the Punjab province are summarized in the following table.

| Month | Mean Monthly Temperature (lowest value in Punjab) in $^0C$ | Mean Monthly Temperature (highest value in Punjab) in $^0C$ |
|---|---|---|
| January | 2 | 14 |
| February | 2 | 18 |
| March | 7 | 23 |
| April | 12 | 29 |
| May | 16 | 34 |
| June | 20 | 36 |
| July | 18 | 35 |
| August | 18 | 33 |
| September | 16 | 32 |
| October | 13 | 27 |
| November | 9 | 21 |
| December | 4 | 16 |

Table 1.2: Seasonal variation of mean temperature in the Punjab province

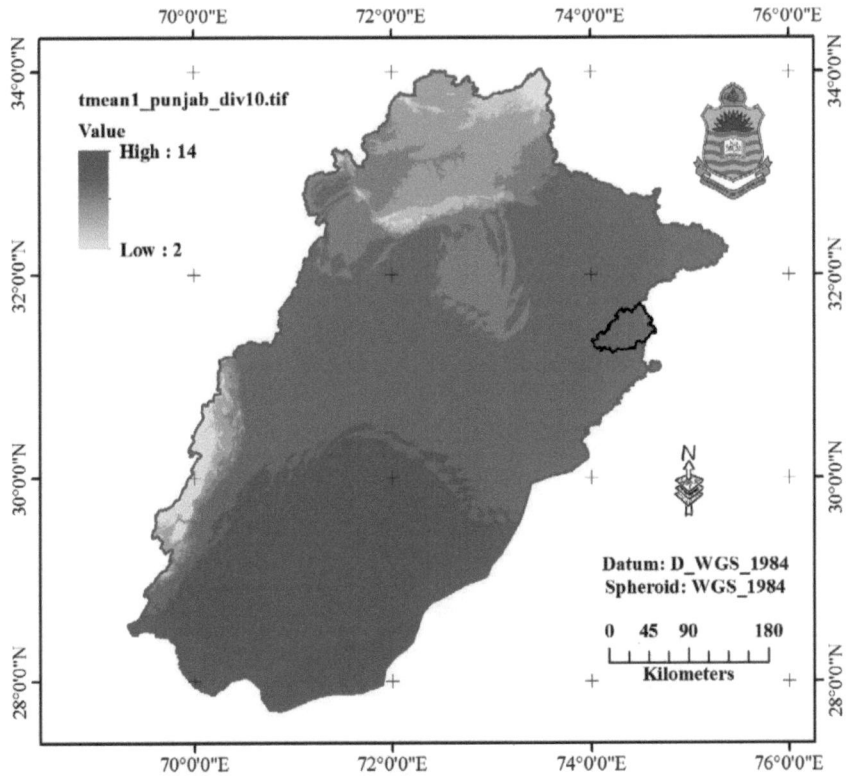

Figure 1.16: Spatial distribution of mean monthly temperature (in degrees Celsius) for the month of January for Punjab province

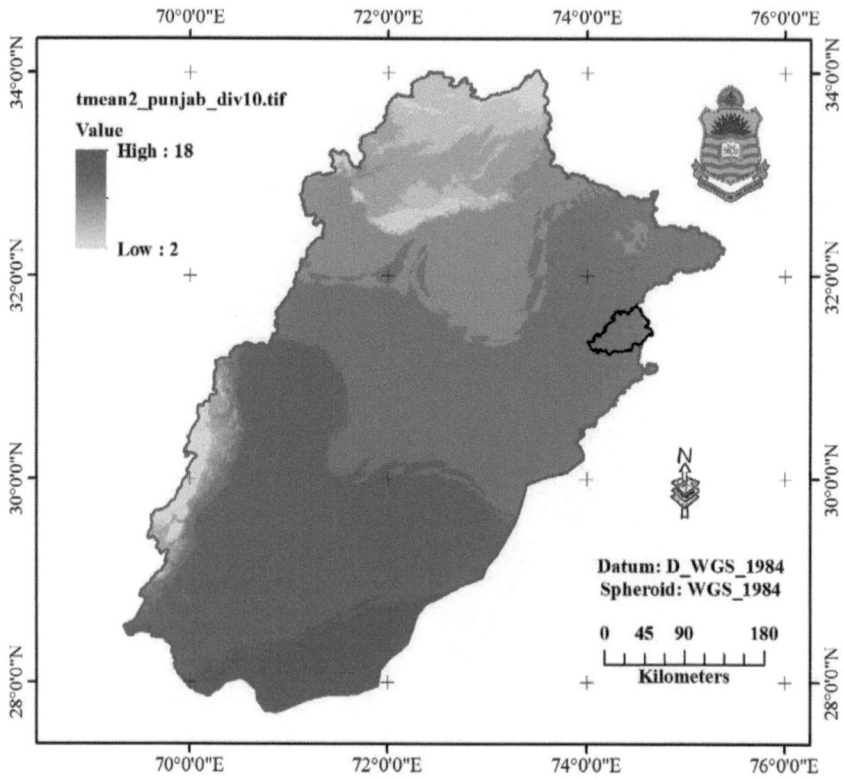

Figure 1.17: Spatial distribution of mean monthly temperature (in degrees Celsius) for the month of February for Punjab province

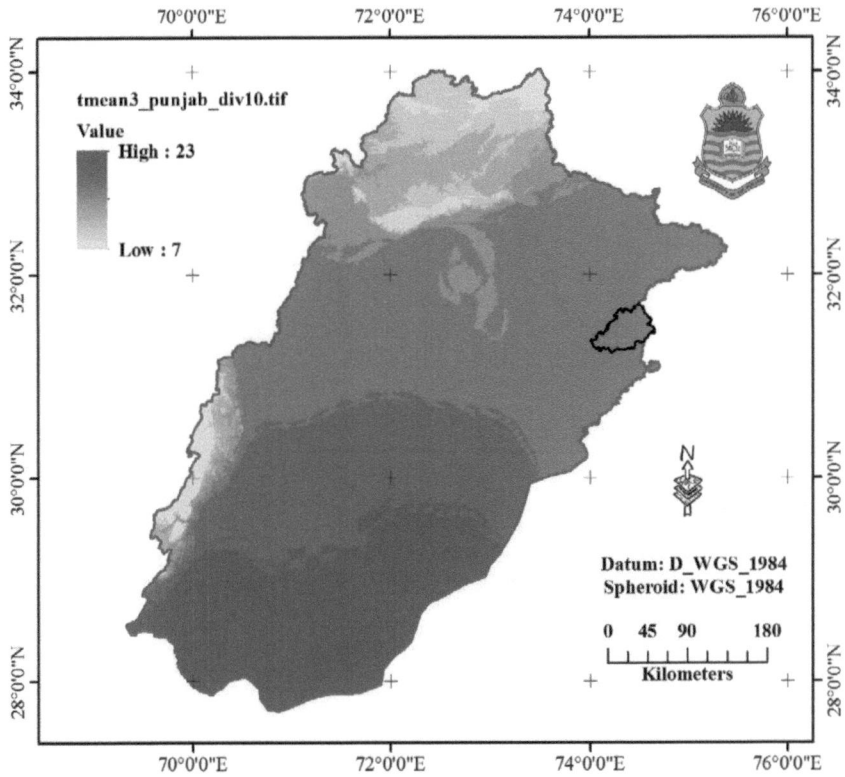

Figure 1.18: Spatial distribution of mean monthly temperature (in degrees Celsius) for the month of March for Punjab province

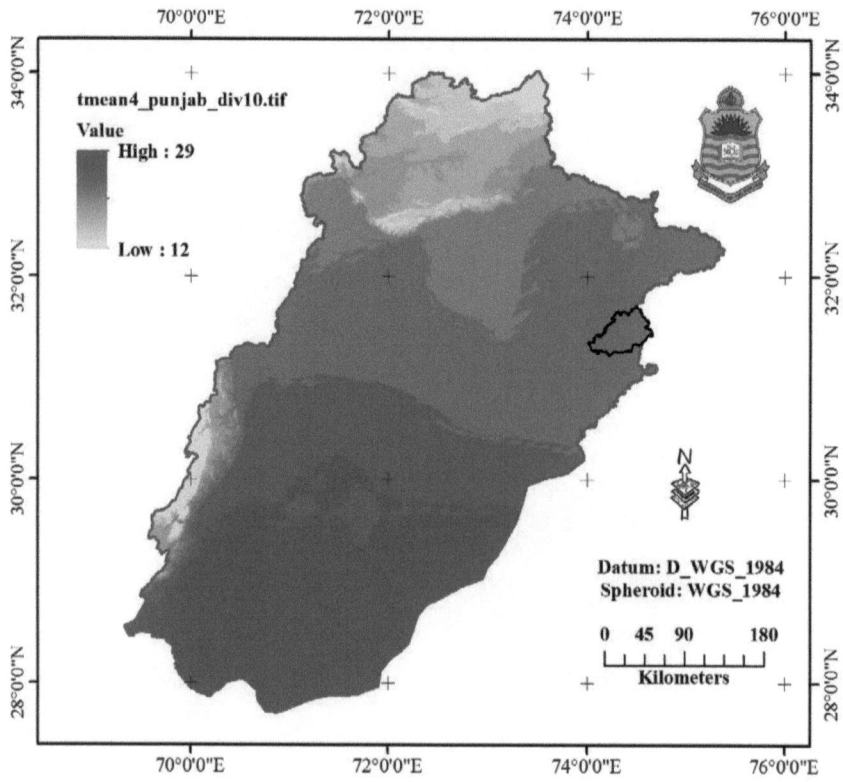

Figure 1.19: Spatial distribution of mean monthly temperature (in degrees Celsius) for the month of April for Punjab province

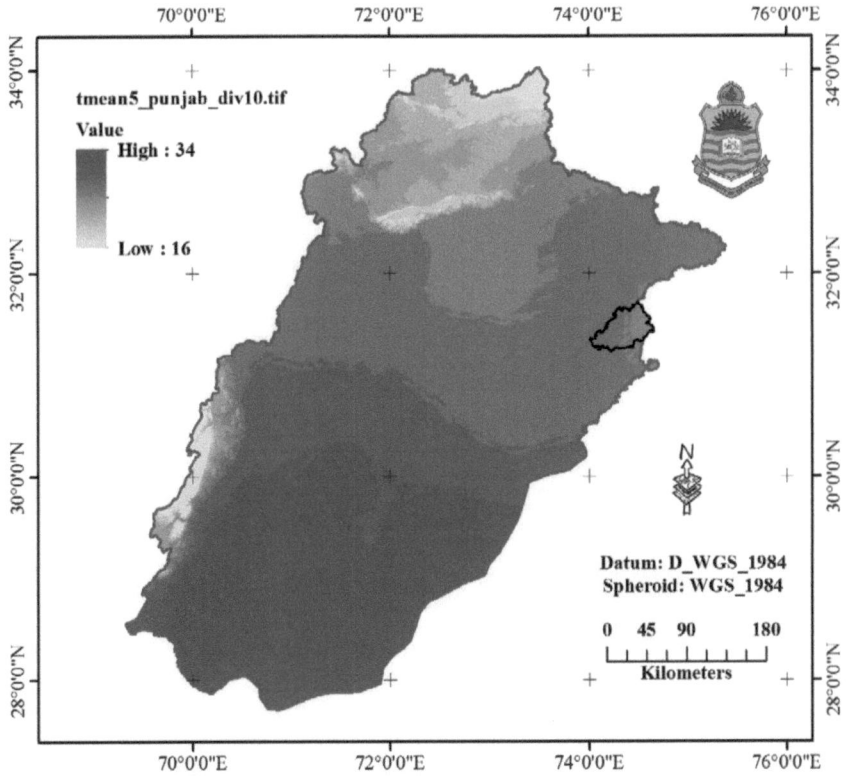

Figure 1.20: Spatial distribution of mean monthly temperature (in degrees Celsius) for the month of May for Punjab province

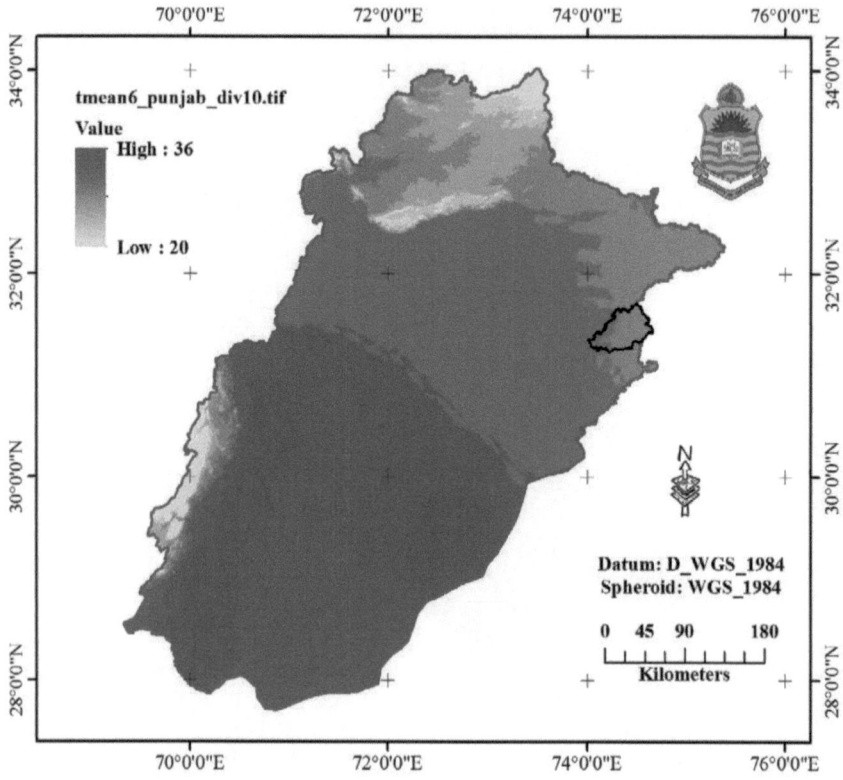

Figure 1.21: Spatial distribution of mean monthly temperature (in degrees Celsius) for the month of June for Punjab province

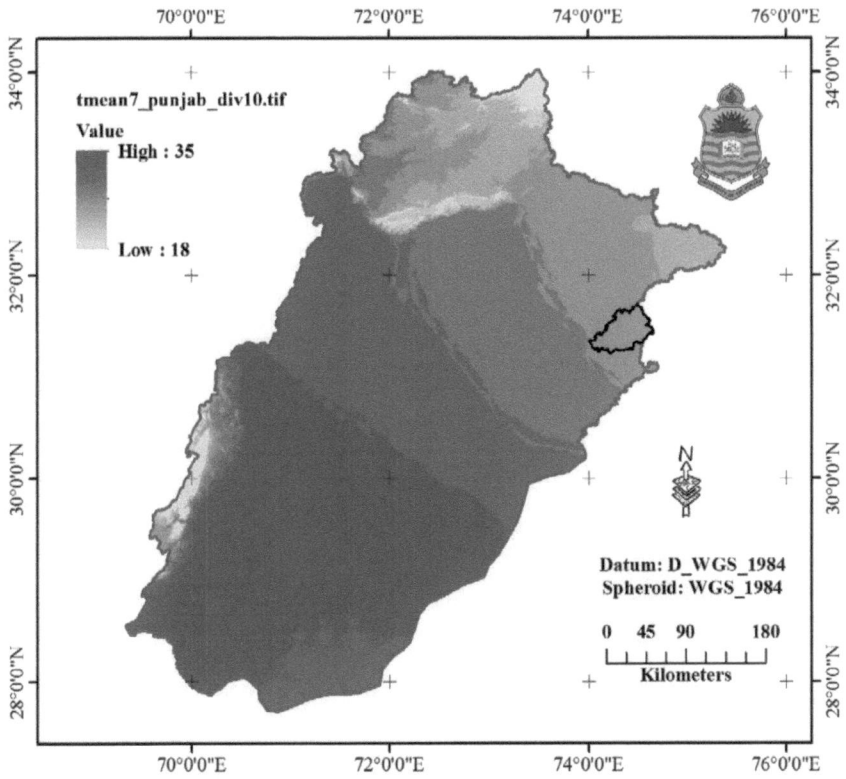

Figure 1.22: Spatial distribution of mean monthly temperature (in degrees Celsius) for the month of July for Punjab province

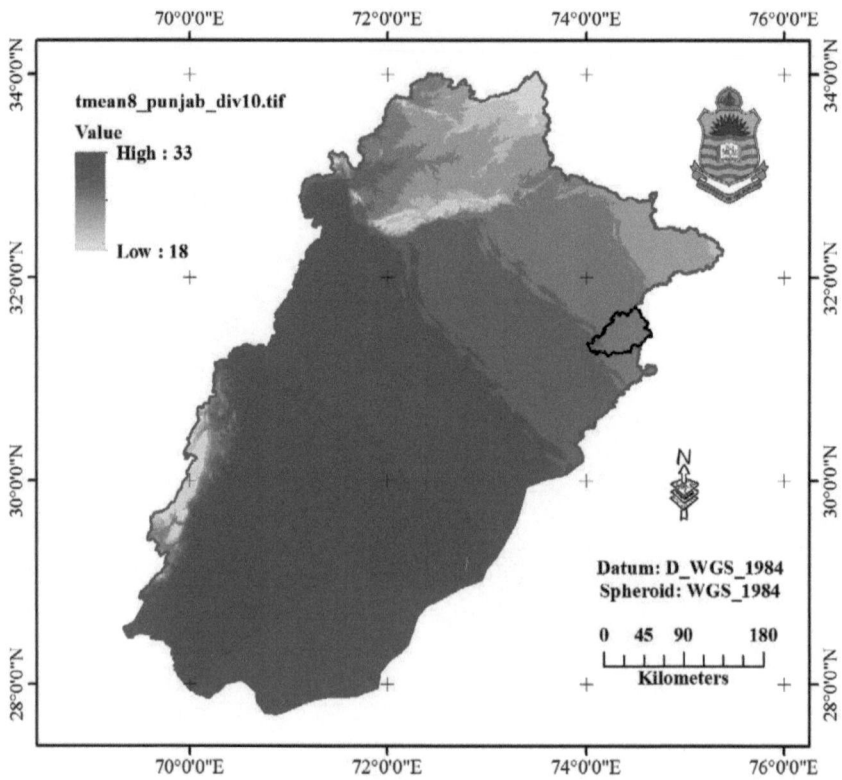

Figure 1.23: Spatial distribution of mean monthly temperature (in degrees Celsius) for the month of August for Punjab province

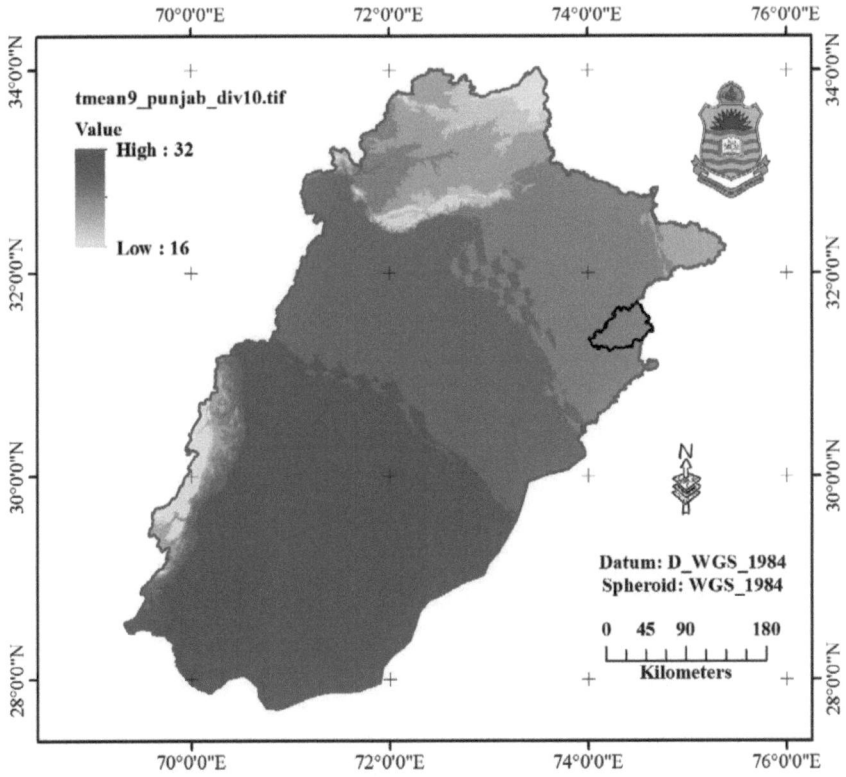

Figure 1.24: Spatial distribution of mean monthly temperature (in degrees Celsius) for the month of September for Punjab province

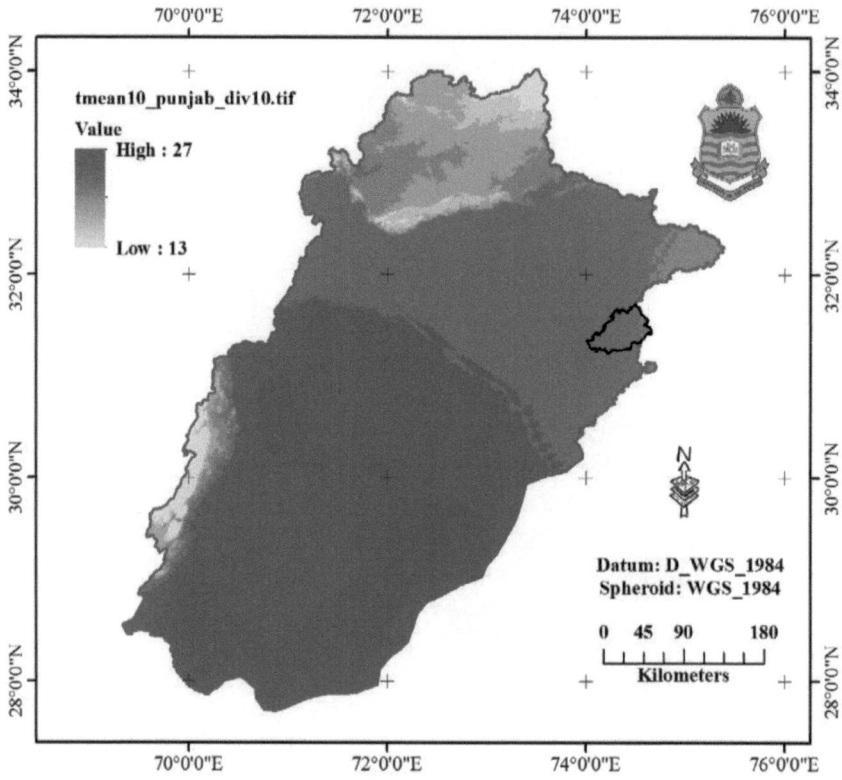

Figure 1.25: Spatial distribution of mean monthly temperature (in degrees Celsius) for the month of October for Punjab province

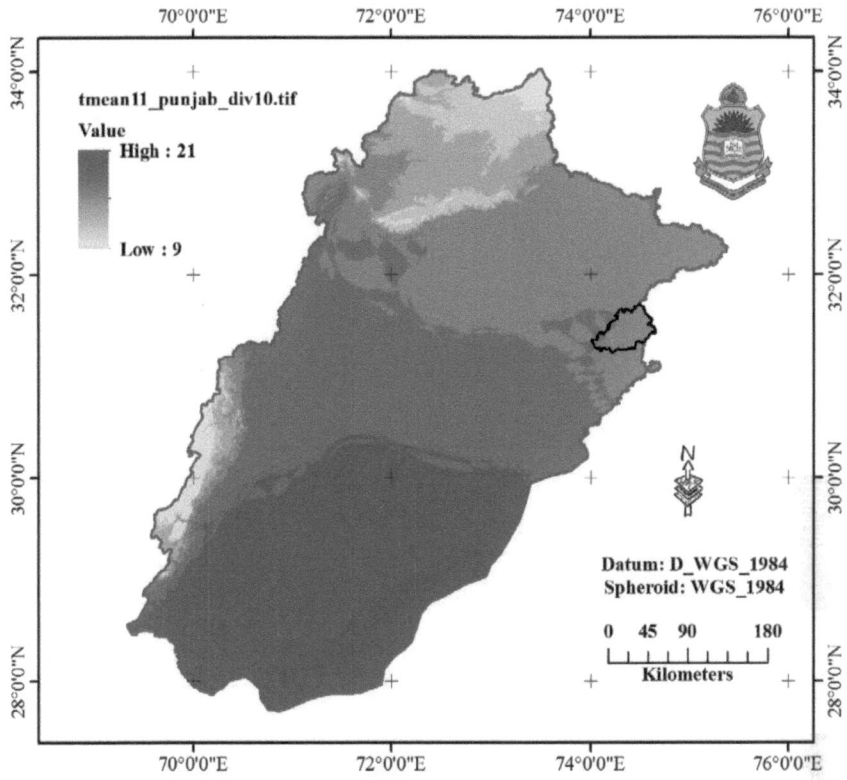

Figure 1.26: Spatial distribution of mean monthly temperature (in degrees Celsius) for the month of November for Punjab province

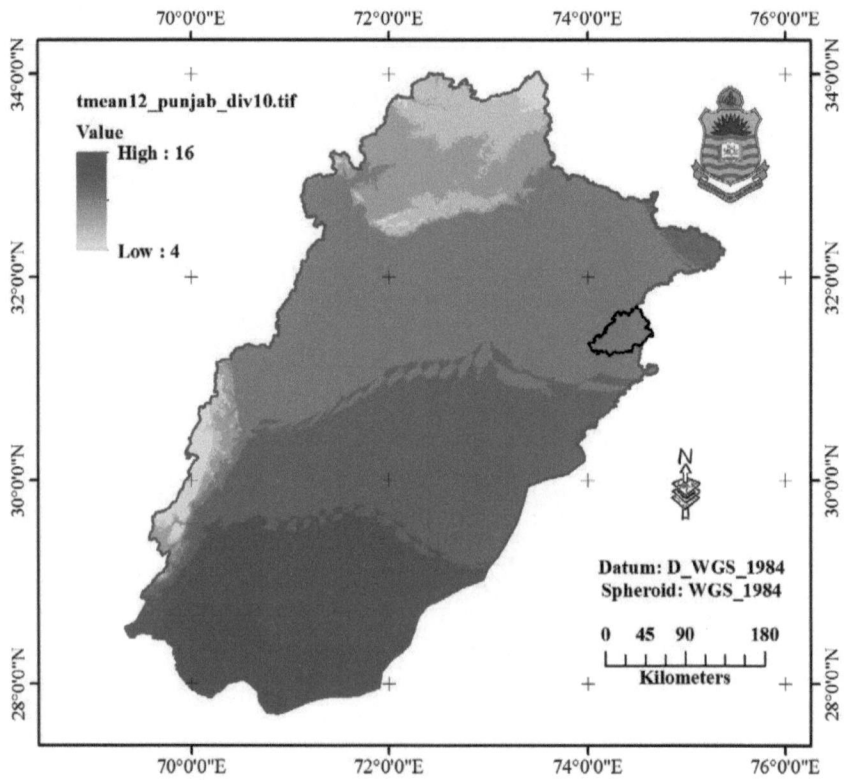

Figure 1.27: Spatial distribution of mean monthly temperature (in degrees Celsius) for the month of December for Punjab province

### 1.2.3 Humidity

The climate of Lahore city is characterised by dry days with low humidity. The days in the monsoonal season bring very high humidity, especially on the day following rain; high humidity is experienced in Lahore. The weather conditions in these days feel very uncomfortable.

### 1.2.4 Wind Direction

The wind direction in Lahore changes from day to day and season to season. However, on the average there is a change of general direction from winter to summer when there is a complete reversal of the prevailing winds from November to May, the predominant wind direction is from West to East. During September and October, wind is changing its direction and is in the state of flux. By and large, it may be said that the predominant wind direction in Lahore is north-west.

## 1.3 Population Growth of Lahore

Lahore is densely populated city. The growing population is exerting more and more pressure on the transportation of the city and creating severe problem of air pollution. Population of the Lahore city has been increasing rapidly from 1951-1998 (Mazhar & Jamal, 2009). Estimates of population for the year 1991 and 1998 also show the same trend. The census of 1998 report a city population of 6,318,745 people living in Lahore with 52.68% male and 47.32% females.

The first few years after the independence, in spite of large number of migrants from India and a significant increase in the population, did not see much of the development in Lahore. The economic development of the recent years has brought about acceleration in the rate of spatial expansion of the city. Lahore has grown almost 14 times in size since the beginning of the 20th century.

## 1.4 Existing Land Use

Due to urban sprawl more and more area of the land is being used for residential purposes and hence the built-up class in the land-use land cover has increased significantly. People residing in the cities put land to different uses. The requirement of our study is to show that how the patterns of air pollution vary in different zones of the city. The commercial centre of Lahore is located along the Mall road. New commercial areas are mixed with the planned residential areas in the Southern part of the UBD canal. Mostly large factories are located along trunk roads in the suburbs and concentrated along Sheikhupura road and G.T. road. According to 1998 census report released by the Pakistan Bureau of Statistics there are 881,708 housing units in the city. There are 150 union councils, 6 towns and 261 mauzas in the district. The population density of the Lahore city is 3565.9 per square kilometres.

There are a number of steel bar plants located north of the Badami Bagh bus terminal and car body manufacturing plants along band road and Multan road. Traditional houses are located in and around the walled city. A mixed use industrial and residential area is situated along the western portion of the Band road. Well-planed housing areas are main Gulberg, Model Town, and Defense. Low-density residential areas are located in the cantonment. Parks, for example; Minar-e-Pakistan and Bagh-e-Jinnah are scattered all over the urban area.

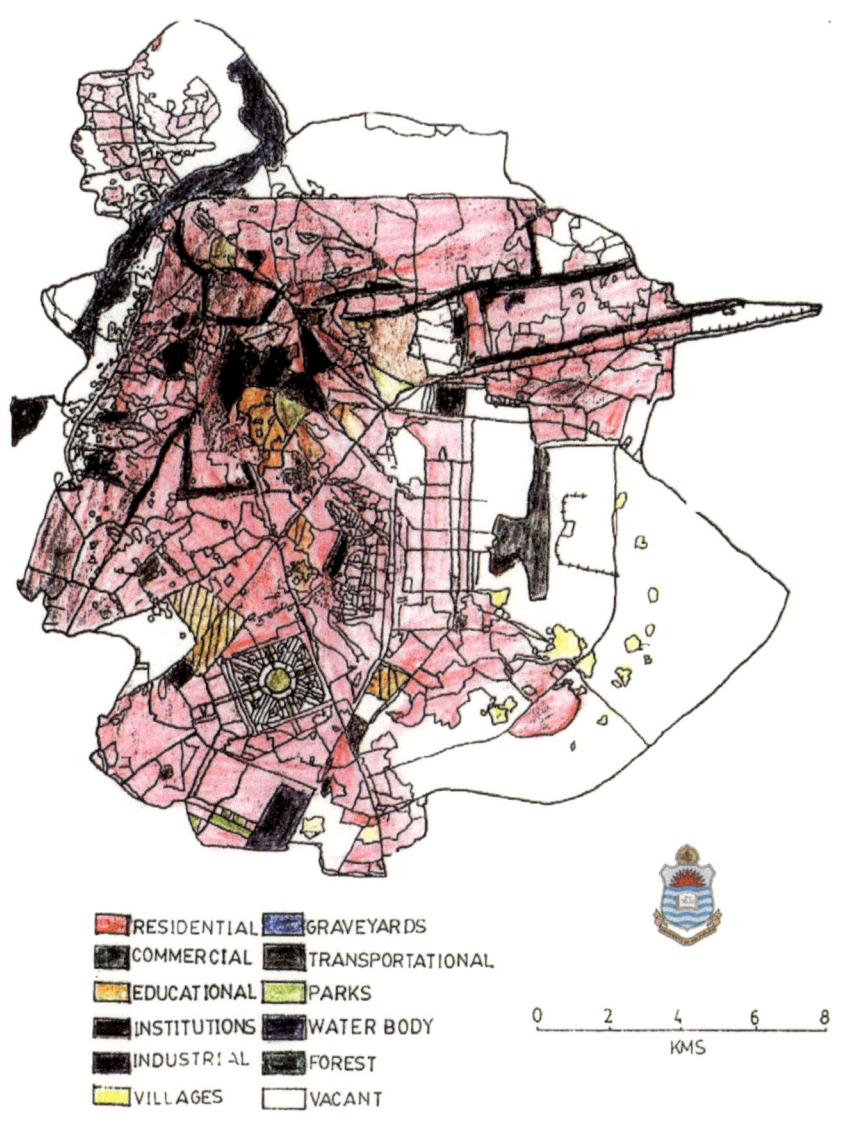

Figure 1.28: Land Use of Lahore City (Isma, 2000)

## 1.5 Air pollution definition

- The term "air pollution" is used to describe substances that are artificially introduced into the air. Air pollution stems from gases and airborne particles, which, in excess, are harmful to human health, buildings and ecosystems. Air pollution worldwide is a growing threat to human health and the natural environment.

- The presence in the outdoor atmosphere of one or more contaminants, such as dust, fumes, gas, mist, smoke or vapour in quantities, of characteristics, and of duration, such as to be injurious to human, plant or animal life or to property, or which unreasonably interferes with the comfortable enjoyment of life and property.

- Air pollution may be described as contamination of the atmosphere by gaseous, liquid or solid wastes or by-products that can endanger life, attack materials and reduce visibility.

- A contaminant that affects human life, plant life, animal life and property or contaminant that interferes with the enjoyment of life and property is an air pollutant.

- Ohio Environmental Protection Agency (EPA) provides a definition of air pollutant or air contaminant as particulate matter, dust, fumes, mist, smoke, vapour or odorous substances.

- When the concentration of a chemical is above the concentration of the chemical present in the air, it is termed as air pollutant.

## 1.6 Common air pollutants

1. Sulphur dioxide
2. Nitrogen oxides
3. Carbon monoxide
4. Hydrogen sulphide
5. Ammonia
6. Chlorine
7. Hydrocarbons
8. Ozone
9. Total suspended particulate matter
10. Lead
11. Carbon dioxide

The air pollutants at serial number 1 to serial number 6 are included in the present study.

## 1.7 Major sources of air pollution

There are two major sources of air pollution which are; 1). the combustion of fuel in stationary sources including the burning of coal, oil, natural gas etc. and 2). production of fuel by-products from its burning in automobiles and transportation. Transportation vehicles produce carbon oxides (CO and $CO_x$) and nitrogen oxides ($NO_x$). Fuel combustion in stationary sources is the main source of sulphur dioxides ($SO_2$).

Here in this study, the main focus has been on the commercial and residential area based air pollutants.

## 1.8 Concerns to air pollution

Air pollution can come in many different shapes and sizes. It can come from many different sources and sometimes change strength over the period of only a few hours. However, though it appears to be such an intangible entity, human could logically prevent up to 65% of all atmospheric pollution.

The problem is that the air pollution is something that we must have to accept if we desire to continue to live a comfortable life. The air pollution is an indirect result of our pursuit of an even higher standard of living. Thus, abruptly ending the 65% of atmospheric pollution that human beings create is what we (as an increasingly industrial society) can unfortunately not afford to do at this time.

### 1.8.1 Should we not worry about air pollution

- Air pollution affects everyone of us
- Air pollution can cause health problems and in an extreme case even death
- Air pollution reduces crop yields and affects animal life
- Air pollution can spoil and corrode materials and monuments
- Air pollution can cause significant economic losses

## 1.9 An insight into Int'l (Asian) air pollution levels

Rapid urbanization, with the associated growth in industry and transportation systems, has increased regional concerns with regard to emissions of sulphur dioxide and nitrogen oxides. The urban growth and, therefore, the increasing rate of energy consumption increase the emissions and could exceed the safer thresholds in the near future. The primary man-made source of sulphur and nitrogen in the Asia – Pacific region is fossil fuel combustion in the energy, industry and transportation sectors. The use of low quality fuel, inefficient methods of energy production and use, the poor conditions of vehicles and traffic congestion are the major causes of increasing emissions of these gases.

Emitted gases have the capacity to be transported over large distances, sometimes many hundreds of kilometres, and may give rise to depositions within the country or in the neighbouring countries. The potential for such transboundary air pollution was evident in the recent Indonesian forest fires (as seen from the newspaper headline). The area affected by their pollutants from the fire spread for more than 3200 kilometres, east to west, covering six Asian countries and affecting around seventy million people. In the Malaysian state of Sarawak, the air pollution index hit record high.

Asia represents a major source of air pollution as a result of rapid population growth, explosive industrialization, and few environmental regulations. Following is a brief description of some countries, which are important in respect of their emissions.

### 1.9.1 Air pollution in China

Due to the use of high sulphur coal to generate energy, the cities in China are heavily polluted by sulphur dioxide and particulate (Streets & Waldhoff, 2000). The average ash content of Chinese coal is 27%. The sulphur content is 5%. The combustion sources include small domestic stoves as well as large industrial plants. China produces over 15 million tons of SO2 and 20 million tons of particulate. Industrial emission of carbon dioxide and greenhouse gases are emitted in large quantities in China. Nitrogen oxide emissions are likely to increase as the production of cars will increase in China. China employs very little air pollution control technology. Acid rain is an important issue in China e.g., Guangzhou and Shanghai are mainly polluted by nitrogen dioxide. The critical pollutant in Beijing was total suspended particle from coal burning, construction projects and dust raising winds.

### 1.9.2 Air pollution in India

Air pollution is serious problem in major cities in India. Delhi's pollution level is highest of all the areas of India and leads to the other areas of the city in high pollution levels. In 23 Indian cities, with populations of more than 1 million, auto exhausts and industrial emissions dangerously cross limits. Recent studies reveal that the number of patients with respiratory diseases and allergies has roughly doubled during 1990 – 2000. In Calcutta, winter levels for particulate matter are 12 times above the standards. In Mumbai's "gas chamber", the eastern neighbourhood of Chembur city, pollution figures zoom to 10 times above the safe levels. India's metropolitan vehicle population has roughly tripled during 1990 – 2000. The most damaging pollutants come from petrol driven cars and two wheelers. Only July 5, 1997, IPAN (Indian Public Affairs Network) published an article of how the growing catalytic convertor used could ease Asian cities air pollution.

### 1.9.3 Air pollution in Indonesia

The air quality in Indonesia is deteriorating rapidly with industrial expansion. In the capital city Jakarta, brownish yellow clouds of lead loaded smoke are common due to the high number congestion of residential areas. During several periods, in 1997, Malaysia experienced haze conditions due to particulate matter from fire burnings in Indonesia. Based upon readings from the Malaysian air pollutant index, the air pollution levels registered in the "unhealthy" and occasionally in the "very unhealthy" ranges.

### 1.9.4 Air pollution in Hong Kong

Vehicle emissions contribute most significantly to Hong Kong's air pollution problems, diesel power engines being the prime reason. Four years ago, in 1996, there were some 300000 vehicles on HK roads, and one in three were diesel vehicles that covered two third of the mileage recorded by the total vehicle fleet. Today (in

2001) there are around 480000 vehicles in the HK, with diesel vehicles accounting for about 60% of overall mileage.

## 1.10 The role of International Cooperation

Although many international environmental agreements may appear inelastic, there are number of reasons why these are useful and should be encouraged. 1). Even a fragile demand means that some action is being taken. 2). The process of preparing international agreements generates information, increases international scientific collaboration and can raise public awareness. According to experienced negotiators, the process of arriving at an agreement can contribute as much to environmental protection as the agreement itself.

# Chapter 2:
# Classification, sources and effects

Pollutants present in the air may be in the form of solids, liquids or gases. These are produced by natural and manmade activities. Naturally occurring processes like dust storms, volcanic eruption etc., create pollution. Man-made activities like industries and civil constructions operations also generate pollution. For convenience of study, various sources of air pollution are classified in different ways.

## 2.1 Type A: Classification

Air pollutants are classified as follows
- Natural sources
- Anthropogenic sources

### 2.1.1 Natural sources (primary and secondary pollutants)

Natural sources include
1. Dust storms are due to local wind circulation, global meteorological processes are responsible for environment with dust pollution
2. Forest fires resulting in huge quantities of smoke and carbon particles
3. Volcanoes, solid particles, gases (sulphur dioxide radiation heat wave's pollution and heavy dust are released)
4. Sea-spray; a continuous phenomenon and a major source of particulate (liquid droplets) pollution in atmosphere

5. Plant pollen spread very vastly due to wind motion, responsible for dust pollution

Primary pollutants are the pollutants which are emitted directly from the sources. These pollutants include sulphur compounds, nitrogen compounds, hydrogen halides and oxides of carbon.

Secondary pollutants are the pollutants which are formed by chemical interaction with primary pollutants. These pollutant include sulphur trioxide, sulphuric acid, ketones, ozone etc.

### 2.1.2 Manmade sources

Manmade sources include domestic pollution generated by house hold activities or infection / using insecticides, industrial pollution, thermal power plants, chemical plants, cement plants, paper mills, textile mills, tanneries as well as traffic pollution because of rapid and unplanned urbanization effects in the form of exhaust gases, particulate and noise.

## 2.2 Type B: Classification

There are three major sources of air pollution. Each type of sources is further classified either as instantaneous sources or continuous sources.

- Point sources
- Line sources
- Area / volume sources

### 2.2.1 Point sources

Instantaneous point source means small volume of material (puff) is released in a relatively short time. Its examples are nuclear explosion and volcanic eruption.

Continuous point source means the polluted gas coming out continuously from a point source (relatively small diameter). For example release of smoke from the stake of power plant.

### 2.2.2 Line sources

Instantaneous line source means polluted dispersion from a line shape source e.g., pesticide spray path emitted by airplanes. The continuous line source means the continuous pollutant dispersion from a line source e.g., urban traffic road.

### 2.2.3 Area / volume sources

Instantaneous area / volume source means pollutant from an area / volume (relatively large comprising with a point source) e.g., blasting of poisonous gas tank etc.

Continuous area / volume source means the pollutants from an area / volume on the continuous basis e.g., heart of urban city complex, area / volume of an industrial park etc.

## 2.3 Sulphur containing compounds

Coal, oil containing sulphur as impurity When fuel is burnt, sulphur also burns to produce sulphur dioxide gas and also sulphur trioxide gas (both are the dominant oxides of sulphur in atmosphere).

### 2.3.1 Sulphur dioxide

Sulphur dioxide is the most important oxide emitted by pollution sources. It is a gas which is colourless, non-flammable, non-explosive, taste sensation at (0.3 – 1.0) level, pungent irritating odour at greater than 3.0 ppm. Sulphur dioxide can irritate the upper respiratory tract, can go deep into lungs. It is moderately soluble in water, that is 11.3 g in 100 ml of water.

$$H_2O + SO_2 \rightarrow H_2SO_3 \text{ (weak acid)}$$

It is partly converted to sulphur trioxide or sulphuric acid and its salts by petrochemical or catalytic processes in the atmosphere. In a polluted atmosphere $SO_2$ reacts photo-chemically or catalytically with other pollutants.

### 2.3.2 Sulphur trioxide

Sulphur trioxide is generally emitted alongwith sulphur dioxide in a concentration of 1 – 5 %. It rapidly combines with moisture to form sulphuric acid.

$$SO_3 + H_2O \rightarrow H_2SO_4$$

Both sulphur dioxide and sulphur trioxide are washed out from the atmosphere by rainfall.

### 2.3.3 Hydrogen sulphide

Hydrogen sulphide is a bad smelling gas produced during an aerobic biological decomposition and also from craft pulp plants etc. Its sources include geothermal power plants, petroleum production and refining. Its effects are severe gas nuisance odour (rotten egg smell), headache and breathing difficulties at higher concentrations. Its control and prevention may be carried out by the control of emissions from geothermal power plants, petroleum production and refining, sewer and sewerage treatment plants.

## 2.4 Nitrogen containing compounds

Nitrogen containing compounds are released by petroleum operations, industrial and automobile combustion. Their primary pollutants are nitric oxide (NO) and nitrogen dioxide ($NO_2$) and are referred as $NO_x$. Generally NO and $NO_2$ are analysed together. There are seven oxides of nitrogen which are $N_2O$, NO, $NO_2$, $NO_3$, $N_2O_3$, $N_2O_4$ and $N_2O_5$. First three namely, nitrous oxide ($N_2O$), nitric oxide (NO), and nitrogen dioxide ($NO_2$) are formed in the atmosphere.

### 2.4.1 Nitrous oxide

Nitrous oxide is a colourless, odourless gas. Its typical concentration in the atmosphere is 0.5 ppm and is released due to biological activity of the soil. Nitrous oxide is called as laughing gas and is anaesthetic.

### 2.4.2 Nitric oxide

Nitric oxide is also a colourless, odourless gas formed under high temperature combustion process. It is not a pollutant and inert at normal temperature conditions. Nitric oxide is emitted by automobiles (in quantities larger than $NO_2$) and is oxidized to $NO_2$ in a polluted atmosphere through photochemical reactions. The high temperature combustion process can be represented by

$$N_2 + O_2 \rightarrow 2NO$$

### 2.4.3 Nitrogen dioxide

Nitrogen dioxide is a brown pungent gas with an irritating odour. $NO_2$ is corrosive to material and is toxic to men. It absorbs sunlight and initiates photochemical reactions that produce smog. $NO_2$ is emitted by fuel combustion and industrial plants which produce nitric acid. It is heavier than air and readily soluble in water to form nitrous acid and nitric oxide. Both combine with ammonia in atmosphere to form ammonium nitrate and fall with rain. Fossil fuel combustion generates nitrogen dioxide and nitric oxide which is rapidly oxidized to $NO_2$. $NO_2$ reacts in the presence of sunlight to form ozone and contributes through atmospheric reactions to the formation of nitrous and nitric acid aerosols. The major sources of nitrogen dioxide include motor vehicles, power plants and other fossil fuel burning industries. Local levels tend to vary with traffic density. Indoor exposures to $NO_2$ can be substantial from unvented combustion sources, such as gas stoves and space heaters. In the absence of indoor sources, indoor levels are about half of those outdoors. The highest ambient one hour exposures reported by EPA are over 0.200 ppm and the highest annual mean exposures are over 0.040 ppm.

## 2.5 Carbon containing compounds

Carbon containing compounds are carbon monoxide and carbon dioxide. These are described below.

### 2.5.1 Carbon monoxide

Carbon monoxide is a primary pollutant. It is a colourless, odourless and tasteless gas. It mainly originates from incomplete combustion of fuels (carbonaceous material) especially due to automobile exhaust. CO is highly poisonous gas (asphyxiant). It is present in small concentration in atmosphere with a life time of 2-4 months. Its toxicity is due to its affinity for haemoglobin hence non-toxic to insects and the life having no red blood cells as well as poses no detrimental effects on material surface and vegetation. Local accumulation in heavy traffic is the most important source for community ambient exposure. Sources of community exposure for CO inside a passenger car are at 5ppm, 15 mm in proximity to busy roads/intersections, 4 ppm in the parking areas, 5-42 ppm inside traffic tunnels. Other important contributors to CO exposure include traffic volume, traffic speed, winter season, motor vehicle density, age composition of the fleet, emission standards for the fleet and vehicle characteristics, community combustion of oil, gasoline, coal, wood and use of lawn movers, chain saws, space heaters and charcoal.

Smokers typically have carboxyhaemoglobin (COHb) levels of 5-6%. Setting side tobacco and indoor sources, the relationship between ambient CO concentrations and blood levels is largely determined by duration of exposure and ventilation rate; the latter is roughly correlated with workload. Wintertime COHb in a normal adult is about 1.2%, with 3-4% of population above 2%. For example, during heavy labour in a busy traffic tunnel with CO levels of 42 ppm, the COHb would reach about 5% in 90 minutes. Risk groups include commuters, smokers, and persons working in traffic. Persons with cardiac and pulmonary disease are most vulnerable. Symptoms such as dyspnoea and angina may develop at COHb bevels of 3-4%.

### 2.5.2 Carbon dioxide

Carbon dioxide is produced with CO but it is not considered an air pollutant. It is produced by burning of fossil fuels such as coal, oil and natural gas and accumulates over the earth for the past five hundred years.

Manmade emissions are more than the natural cycle which is represented below

$$CO_2 \rightarrow \text{Air/ Dissolved in Water} \rightarrow \text{Plants} \rightarrow \text{Animals} \rightarrow CO_2$$

The earth surface temperature is directly related to $CO_2$ in the atmosphere because carbon dioxide strongly absorbs long wave (infrared terrestrial radiation and continuous $CO_2$ build up, trap heat and prevents some of it reradiate it back into space. Carbon dioxide is uniformly distributed upto 75km vertically over the earth surface. About half of the extra $CO_2$ produced by combustion of hydrocarbons is retained in the atmosphere.

## 2.6 Residential (indoor) air pollution sources

There are five types of pollution sources in the home. The first to be recognized was burning fuel indoors for cooking and space heating. Natural gas, the most commonly used indoors mainly produces nitrogen dioxide and carbon monoxide along with harmless combustion products. If wood is burned on a fireplace or for cooking food then in addition to these two pollutants, particulate matter and a host of potentially hazardous hydrocarbons are produced. Burning coal or oil produces all of these pollutants plus sulphur dioxide. Because most of the oil heated homes use a low sulphur grade of oil and dew houses use coal therefor in this particular case sulphur is not a major problem. This problem is associated with the use of coal at homes, which is widely used in many countries like China. In developing nations where poorly vented stoves and fireplaces are common, indoor pollution from fuel burning is believed to be a major source of health hazard.

A second source of indoor air pollution results from synthetic and some natural materials used for carpeting, foam insulation, wall coverings and furniture. Glue used in some plywood, for example, fives off formaldehyde. Asbestos used as building material because of its heat-resistant properties can put asbestos fibre into the indoor air if the material is not properly sealed. In offices some copy machines and computer printers are a source of toxic organic substances like toluene.

Indeed the air in many modern office buildings is particularly polluted because of the combinations of office equipment, synthetic carpets and poor ventilation. In this regard, there should be some safety regulations in industry to protect workers against these affects (headache and eye irritation are common as a result of poor air indoor air quality). A third source of indoor air pollution is toxic gases leaking upward into the living area of a home from the soil beneath the house. Recent evidence suggest that toxic gases may probably enter houses located near chemical factories and waste dumps this way as well, although the extent of this toxic gas penetration is not really known. Many commercial products such as furniture polish, glues, cleaning agents, cosmetic deodorizers and solvent used in home contribute to the toxicity of indoor air. Moreover, dry cleaned clothes brought into the home are a source of chlorine containing compounds.

The fifth source is tobacco smoking. Not only it is a serious indoor air pollutant in its own right, but also contains carbon monoxide that is injurious for human health. Tobacco smoking or living with a smoker increases the risk of illness from other toxic compounds.

## 2.7 Effects of air pollution

### 2.7.1 Effects of air pollution on atmospheric properties

Air pollution affects the atmospheric properties in a mercurial way. It reduces the visibility and cause traffic hazards. A visibility of 50-60 km is possible in rural setting i.e., pollutant concentration of upto 20 micrograms per cubic metre. It reduces to 8-10 km in urban setting where 100 micrograms pollutant concentration is present in one cubic metre of air. A typical setting of moderate fog gives 5-7 km visibility whereas heavy pollution gives visibility of only 2-3 km. Three major effects caused by air pollution are reduction of solar radiation, formation of heat islands or greenhouse effect, fogging and precipitation. Urban areas as compared to rural situation receive around 20% less solar radiation which is further reduced by upto 10% during heavy pollution. Greenhouse effect (heat island) is caused by concrete jungles built in urban areas having a high specific heat, these built-up areas absorb more solar radiation / energy and reflect less amount back to atmosphere. Pollutants such as sulphur dioxide in sulphurous smog and ozone in photochemical smog cause greenhouse effect, particularly at night hence pollutants increase incidence of fog formation. High concentrations of sulphur dioxide lead to formation of $H_2SO_4$ under humid environment thus resulting in the formation of fog droplets. These fog droplets cause eye irritation and temporary / permanent damage to the respiratory system.

## 2.7.2 Effects of air pollution on human health

The effects of air pollution on human health generally occur as a result of contact of air pollutant and human body. These contacts occur at the surface of skin and exposed membranes. The most important is the contact with the exposed membrane, because of their high absorptive capacity. Airborne gases, vapours, fumes, mist and dust may cause irritation of eye membrane, nose, throat and lungs. Eyes and respiratory system are the major affected areas.

The respiratory system consists of lungs and respiratory tract for air passage from nasal cavity to the lungs. The function of respiratory system is to inhale air in to the lungs, to filter the impurity from the inhaled air, to supply oxygen to the blood circulatory system and exhaled carbon dioxide from the blood. Particulate matter can deposit in various regions of the respiratory system depending upon the size of particle.

Eye irritation may result when gaseous or particulate material comes and contact with external coat of the eye and with the internal mucus line of the eyelid. Irritation of the eye leads to rubbing which may cause physical damage.

Inorganic gases cause hazardous effects on health. The inorganic gases include carbon monoxide, sulphur dioxide and nitrogen oxides.

When carbon monoxide is inhaled, it passes through lungs and defuses into blood. CO gas has a great affinity to combining with the haemoglobin, which is present in the blood, this combination leads to the formation of carboxyhaemoglobin. The affinity of haemoglobin to absorb carbon monoxide then increases and is 200 times more than to absorb oxygen. Due to lack of oxygen (asphyxiation) death can occur. The ambient air quality standard is 10 ppm for 8 hours. At 10ppm, most of people experience dizziness, headache. Cigarette smoke contain 400-450 ppm of CO. Death can occur for a short period of exposure at concentration of higher than 750 ppm. People feel less consciousness at 250 ppm. Urban busy traffic areas are at level 5-20 ppm.

Sulphur dioxide can cause irritation reduction of visibility and respiratory diseases. Healthy person may experience bronchi – congestion at 1.6 ppm of $SO_2$ for a few minute of exposure. Throat irritation occurs at 8-12 ppm level. 10 ppm can cause eye irritation. 20 ppm results in immediate cough and eye irritation. Exposure of 400-500 ppm (even for a few minutes) is dangerous to life. Normal urban concentration level is 0.001 to 0.200 ppm.

Nitrogen oxides (including NO and $NO_2$) are of interest to the concern of human health. NO is not irritant and it will cause no adverse effects at atmospheric concentration, however, it pose health hazard as an oxidant when it undergoes oxidation to $NO_2$. Haemoglobin has 300000 times affinity for absorbing $NO_2$ than $O_2$, which reduces oxygen carrying capacity of blood. Nitrogen oxides cause lung tissues to become leathery and brittle and may cause lung cancer and emphysema. The safe concentration threshold value of $NO_2$ is upto 0.12 ppm. A concentration of 2-5 ppm for one hour exposure increases air resistance whereas at 150 ppm and above are fatal to health.

The impact of air pollutants on human health is caused by factors such as nature of the pollutant, concentration of the pollutant in the air, duration of the exposure, state of the health of receptor and age group of receptor etc.

The health effects and diseases include lung cancer, chronic bronchitis, bronchioles asthma, emphysema, eye irritation, nose and throat irritation, respiratory tract irritation, increase in mortality and morbidity rate.

Lung cancer is the destruction of lung tissues. It may be caused due to the organic carcinogens like arsenic, asbestos, cadmium, chromium, nickels and radioactive material.

Chronic bronchitis is the reduction of efficiency of air delivery by reducing the diameter of bronchitis. Bronchioles asthma is the more responsive of the trachea and the bronchi to various stimuli, which will occur by narrowing of airways.

Emphysema is due to progressiveness of the breakdown of alveolar air sacs in the lungs by chronic infection or irritation of the bronchiole tube, paralysis of cilia, and injury from violent coughing. Emphysema disables the ability of lungs to exchange oxygen and carbon dioxide to bloodstream.

Hydrogen fluoride causes diseases of bones (fluorosis) and mottling of teeth. Respiratory diseases like siliceous and asbestosis are caused by the dust particles.

### 2.7.3 Airborne diseases

Airborne diseases are those which are transmitted by air. These includes the following
- Respiratory diseases such as pulmonary tuberculosis (bacteria), influenza (viral infection) and pulmonary mycosis (jungal) are transmitted by air
- Coughing and sneezing may be produced due to the aerosols or fine droplets that contain microorganisms (pathogens), from the lungs, leave the infected part via mouth and nose
- Airborne infections are those transmitted by pathogens carried on the very small droplet nuclei to be taken into the body via the respiratory system
- Small particles can penetrate to the lower respiratory system. Large particles are removed by defence mechanism

- Pulmonary tuberculosis (TB) is the most significant airborne disease. It remains one of the main causes of the disability and death throughout the world
- According to the WHO, about 2-3 million deaths per year occur due to air borne disease

### 2.7.4 Air pollution effects on the materials

- The damage caused by atmosphere pollutant to material is a well-known phenomenon
- Particulate such as soot, fumes, soil can cause damage to exposed surface and fabrics because of their abrasive nature
- $SO_2$ is the most notorious pollutant responsible for the metallic erosion
- It has been estimated that erosion of hard metal (like steel) begins at annual mean concentration of 0.02 ppm (52 microgram per cubic centimetre)
- At level of 0.09 – 1.00 ppm, $SO_2$ affects fabrics, leather and paint
- $SO_2$ is readily absorbed by leather and causes its disintegration
- Paper is also disclose by $SO_2$ and becomes brittle and fragile
- $H_2SO_4$ mist in atmosphere causes deterioration of structural material e.g., marble and limestone. The building of Taj Mehal is also under the influence of $SO_2$
- $O_3$ is very reactive substance so much of degradation of material such as cracking and weakening of rubber and fabrics are due to weathering and oxidizing of ozone
- Ozone causes the cracking of synthetic rubber at atmospheric concentration level of 0.001 0.002 ppm (20 – 40 microgram per cubic centimetre)
- It also attacks fabric fibre and the adverse effects increase in the order that fibre of cotton first and acetate nylon and polyesters later
- The fading of fibre and the cracking are attributed to the ozone oxidizing ability

- Nitrogen oxides are known to cause fading the air acetates, cotton, rayon fibre at level of 0-2 ppm over 2-3 months exposure
- Particulate nitrate attacks and damages nickel brass alloy in the presence of moisture

## 2.8 Effects on plants

There are three kinds of injuries to plants caused by air pollutants. These are acute injury, chronic injury and growth / yield retardation.

Acute injury results from short time exposure to relatively high concentrations, such as might occur under fumigation conditions. The effects are noted within a few hours to a few days and may result in visible markings on the leaves due to collapse and death of cells. This lead to neurotic patterns i.e., area of dead tissues.

Chronic injury results from long term low level exposure and causes chlorosis or leaves abscission.

- The degree of opening of stomata controls the gas absorption by the leaves
- When stomata are wide open, absorption is maximum
- A plant is subject to injury by absorbing a pollutant gas like $SO_2$
- High light intensity of morning, high relative humidity adequate moisture supply to the plants roots moderate temperature is the factor to cause the stomata opens
- Most plants close the stomata at night are less sensitive than at daytime

Growth or yield retardation is the injury in the form of effects on growth without visible mark. While deciding about the type of injury the plant has suffered one must carefully distinguish between the adverse effects caused by the air pollution and adverse effects resulting from each factor. Plant disease may show symptoms, which are very similar to these caused by air pollution high temperatures, poor plant care, shortage of nutrients and water may also cause an appearance, which is similar to that of plant damage by air pollution. It may also be due to insect damage. Therefore, while diagnosing the effects of air pollution, one must consider factors such as plant disease, mutation history, weather damage and the nature of pollutant in the area.

## 2.9 Types of plants in Lahore

The climate if the Lahore city is characterized by hot semi-arid air. The summer season is extremely hot, rainy, and longer. The winters, however, are cold and dry. The monsoon rainfall occurs during both summer and winter seasons and dust storms during summers.

Most common crops in the area are wheat, rice, sugarcane and fodders. There are a number of various types of plants in the Lahore city. Common trees of Lahore include: Alstonia scholaris (locally termed ditabark), Bombax malabaricum (locally termed sunbal or silk cotton tree), Callistemon citrinus (locally termed bottle brush), Dalbergia sissoo (locally termed shisham), Delonix regia (locally termed gulmohar), Erythrina suberosa (locally termed coral or gul nister), Ficus benghalensis (locally termed banyan), Ficus religiosa (locally termed pipal), Ficus retusa (locally termed bobari), Kigelia pinnata (locally termed gul-e-fanoos or sausage), Livistona chinensis (locally termed bottle palm), Mangifera indica (locally termed aam), Mimusops elengi (locally termed molsery), Pongamia pinnata (locally termed sukh chayn or Indian beech), Syzygium cumini (locally termed jamu), Ziziphus zizyphus (locally termed jujube).

### 2.9.1 Roadside Plants

There are several roadside plants in Lahore which could survive and grow in the polluted environment. These include *Fraxinus americana* L., *Platanus acerifoliathe, Azadirachta indica, Cynodon dactylon, Calotropis procera, Cenchrus ciliaris, Heteropogon contortus, Millettia thonningii* and *Alstonia scholaris* etc. The *Alstonia scholaris* is the most common roadside plant found in many locations in Lahore. While studying the impacts of air pollution on plants, there are several parameters on which the stress on plant can be measured. These parameters include the amount of dust, percentage of leaf moisture content, photosynthetic rate, transpiration rate, stomatal conductance, chlorophyll content and the quantity of carotenoids in the leaf samples.

### 2.9.2 Ediphytes and Chasmophytes

The term Ediphytes is used to describe the plants grown on the buildings, walls and or concrete structures. These are vascular plants growing out from the moist wall fissures, gaps and cracks caused by leaking water from high rise buildings. Ediphytes are an important in urban areas. The roots of Ediphytes grow deep inside the concrete structure and they create damage to the buildings. As a result of this destructive capability, the historical buildings become endangered. Whereas the term "Chasmophyte" is used to describe the plants growing in abandoned places as well as damaged foundations of old structures. The Chasmophytes are known to be stress tolerant in nature and hence are present in the disturbed areas. Some important botanical families (and their species) present in Lahore are Asteraceae (Xanthium strumarium L., Eclipta alba L., Conyza Canadensis L., Parthenium hysterophorous L.), Brassicaceae (Coronopus didymus, Farsettia jaquemontii), Caesalpiniaceae (Cassia occidentalis L.), Chenopodiaceae (Chenopodium album L., Chenopodium murale L.), Commelinaceae (Commelina benghalensis L.), Cyperaceae (Cyperus rotundus L.), Euphorbiaceae (Euphorbia prostrata Ait., Euphorbia indica L., Euphorbia hirta L.).

## 2.10 Sensitivity of the plants to air pollution

The sensitivity of the plant to air pollution is conditioned by many factors. These factors include generic factors, climatic factors and some miscellaneous factors such as soil, water and fertility effects on the sensitivity of plants to air pollutants.

Plant response to pollutant varies between species of a given genus and between variable within a given species. Such variation is the function of generic variability as it affects the plant morphological, physiological and biochemical characteristics.

The important climatic factors affecting the response of vegetation to air pollutants are duration of light, light quality (wave length), light intensity, temperature and humidity.

### 2.10.1 Damage to Leaf Structure

Chlorosis is caused in the leaf which is apparently visible by the yellowing of the plant leafs. The leaf structure is damaged due to the insufficient values of Chlorophyll which is essential for photosynthesis. Exposure of plant leaf to high concentration of air pollutants such as ozone etc., could cease the growth process in plants.

### 2.10.2 Delayed Flowering

Delayed flowering process occurs especially in the roadside plants. Scientific studies have shown that plant's exposure to high level of vehicle's emissions damages the plant structure and plant utilize its energy in its survival which significantly delays the flowering

### 2.10.3    Root Damage

Emissions of $SO_2$ in the atmosphere could mix up with the rainfall to become acid rain. The soil become acidic which damages the root system and prevent the uptake of nutrients from soil required for plant growth. Moreover, the plant bacteria are reduced as a result of acidic soil which reduces the soil's micro-organisms. Scientific studies indicate that the root length of some specific road side plants such as *Peltophorum pterocarpum* and *Leucaena leucocephala* are more affected with the toxic effects of lead and cadmium.

## 2.11 Residence time of air pollutants

The concentration of an air pollutant in the air at an instant of time depends on the level of turbulence and atmospheric motions. The pollutants do not stay in the atmosphere for ever. These are cleaned out of atmosphere by various physical processes. These processes include dry deposition, wet deposition (together with the falling precipitation), and their conversion to other forms by absorption as well as oxidation.

The resident time is the average amount of time the pollutant spend in the air. Their residence time in the air depends on the source characteristics, atmospheric diffusion processes and sink mechanisms (Bolin B. et al., 1973). The sink mechanisms for some air pollutants are described in the upcoming paragraph in this chapter. Typical residence time of major air pollutants are given below in Table 2.1.

Sulphur dioxide is removed from the atmosphere in about 40 days. In the atmosphere it combines with $O_2$ to form $SO_3$ and then $H_2SO_4$ as well as Ammonia Sulphate. Hydrogen Sulphide stays in the atmosphere for about more than 50 hours. It is mainly produced from biological decay. $H_2S$ is cleaned out of atmosphere by chemical reactions by forming insoluble metal sulphides.

| Name of Pollutant | Approximate Residence Time |
|---|---|
| $N_2$ | 106 year |
| $O_2$ | 10 year |
| $CO_2$ | 15 year |
| $CH_4$ | 10 year |
| $H_2$ | 10 year |
| $N_2O$ | 150 year |
| CO | 65 days |
| $NH_3$ | 20 days |
| $NO/NO_2$ | 1 day |
| $O_3$ (troposphere) | < 1 year |
| $HNO_3$ | 1 day |
| CFC | 65 year |
| $SO_2$ | 40 days |
| $CH_3CH$ | 3 - 13 hours |
| $(CH_3)S$ | 31 hours |
| $H_2S$ | 53 hours |

Table 2.1: Residence time for different pollutants in atmosphere (Esmen, et. al., 1971)

Ammonia is cleaned out of atmosphere by reactions with acids. It forms oxides. Fluorides are very reactive to react with organic compounds and carbonates. Oxides of nitrogen react to form nitric acid and calcium nitrate by reacting with ammonia or lime. Hydrocarbons are cleaned by photochemical reactions. Nonreactive Hydrocarbons are removed by biological reactions. Carbon monoxide is removed by soil absorption and chemical oxidation. Carbon dioxide is removed by biogenic processes.

Wet deposition involves rainfall, snowfall and dew. Very small particles (less than 2 micron are not deposited by rainfall and could stay in the atmosphere for longer periods of time. These settle down very slowly.

The rate of dry deposition of an air particle depends on the deposition velocity ($V_D$) and pollution concentration. The deposition velocity is determined by the specific factors such as pollutant's solubility, particle's diameter and density, wetness/roughness of surface, wind speed and turbulence. Common values of deposition velocities of some common air pollutants is given in Table 2.2.

| Name of Pollutant | $V_D$ (cm/s) |
|---|---|
| $O_3$ | 0.2 to 0.7 |
| NO | 0.01 to 0.1 |
| $NO_2$ | 0.1 to 0.8 |
| $MNO_3$ | 0.5 to 5.0 |
| $NH_3$ | 0.2 to 0.6 |
| $SO_2$ | 0.2 to 3.0 |
| $H_2S$ | 0.2 to 0.4 |

Table 2.2: Typical deposition velocities of some air pollutant (Source: Air Pollution Research Group of http://www.utoledo.edu/)

# Chapter 3:
# Methods

## 3.1 The air pollution monitoring unit

The low cost air pollution-monitoring unit has been developed indigenously, keeping in mind the environmental objectives to identify toxic gases such as CO, $NH_3$, $CL_2$, $H_2S$, $NO_2$ and $SO_2$ in the ambient air. This process has been established from standard WHO procedures. The development of this unit has been undertaken keeping in view not only recent trends in the field, but also forthcoming innovations.

Each unit contains six kits, which is designed from lightweight plastics and wood making it portable. The box provides maximum safety to chemicals against light and is easy to handle. Each unit contains safe packing of the reagent bottles, the visual comparator and the leaflets of procedure, in addition to volume measuring equipment where required.

### 3.1.1 Equipment requirement

Most of the air sampling in pollution studies is done with some type of vacuum pumping equipment. It is required to draw an air sample through a chamber, which holds a special absorbing solution. The absorbing solution is chemically selective for a particular gas and is held in a special glass tube called "impinge". Impingers are available in various shapes and sizes. The simplest consists of a glass vial plus a stopper with a short and long glass tube. The long tube is immersed into the absorbing solution, which enables the impinge to disperse many minute hubbles. The smaller the bubble, the more surface contact is permitted between the gas and the absorbing solution and higher efficiency of gas absorbing results.

There must be different absorbing solutions for different gases to be detected due to the different chemical behaviour of different pollutants. In order to attain a quantitative test, means to measure the amounts of air drawn through the absorbing solution must be supplied. This is done by the use of adjustable flow meter. This devise measure the rate of airflow through the absorbing solution. For most air pollution studies, flow meter is calibrated separately to measure the rate of flow in litres (l/m). When a flow meter is attached to the vacuum portion of the air-sampling unit, one can accurately monitor the air volume pulled through the absorbing solution.

For outdoor air sampling, the equipment must be battery operated and completely portable.

### 3.1.2 Selection of a sample site

Before the selection criteria are given, it is assumed that the analyst is aware of the background information previously mentioned.

A sampling point is selected if it meets the following requirements:

1). It produces a homogeneous level of pollutants.
2). It is not directly exposed to varying amounts of gas, either through corrective air currents or the movements of other bodies such as traffic, slamming doors or leaking distribution pipes in an industrial unit
3). Day conditions at the sampling point are consistent

4). The point is located where the maximum people are expected to be affected by pollution

### 3.1.3 Setting the sampling point

Once a sampling point has been defined, setting the equipment is easy. The air pump should be placed on a levelled surface (e.g., the ground, a table top, on a deck, in an office or ideally at 4-6 feet height). This should have the impinge tube, containing the proper volume of interacting reagents. The air pump is then switched on for a given period of time according to specific procedures, which are described later.

## 3.2 Objectives and scope of the kit fabrication program

The aim of fabricating air monitoring units is to gather quantitative object estimates of the level of major air pollutants in a particular environment. The unit consists of six kits that allow flexibility in order to accommodate the sampling of various gases. Its aims are 1). to develop technical skills in locally manufactured air-testing equipment, 2). to estimate the current status, extent and fluctuations in the concentration of selected air pollutants, 3). to provide a database to resource managers for long term planning towards pollution control, 4). to create awareness among users regarding air pollution at various sites

## 3.3 The air pollution monitoring unit

The air pollution monitoring unit has been developed keeping in mind the environmental objectives to identify toxic gases such as CO, $NH_3$, $Cl_2$, $H_2S$, $NO_2$ and $SO_2$ in the ambient air. The process has been established from the standard WHO procedures. The development of this unit has been undertaken keeping in view not only recent trends in the field, but also forthcoming innovations.

Each unit consists of six kits, which is designed from light weight plastic and wood, making it portable. The box provides maximum safety to chemicals against light and is easy to handle. Each unit contains safe packing for the reagent bottles, the visual comparator and the leaflets for the procedures, in additional to volume measuring equipment where required.

The air pump is assembled from a locally available motor pump. Their calibration for an efficient flow rate is accomplished using a modification component. The pumps are standardized against a flow meter to give an air discharge rate of 1+0.1 (1/m). Adjustments of the hydrostatic pressure are made via an inlet to outlet effluent air ratio. These pumps have been provided in two forms: one operable on DC dry cells (1.5 V) and the other on 220 AC main supply. The former may be used on locations where electric power may be unavailable and the later on sites where electrical power is supplied. As long as there is a constant voltage supply the durability or output of the unit will not be affected: A frequent change of batteries in the air pumps, (after roughly 20-30 measurements) is recommended for ensuring accuracy in results.

The individual chemical test kits used to analyse the absorbing solution after using the impinger are based upon established methods for testing air pollutants. For convenience and portability these testing units are furnished in compact carrying cases. All of the necessary apparatus and reagents that are needed to conduct a chemical analysis of the air are included within the testing equipment.

The reagent system in the testing units includes an absorbing solution that extracts the pollutants from the atmosphere. In some systems a single indicator is added to the absorbing solutions, which is pre-treated before the indicator is added. The colour reaction is measured with the help of a visual comparator.

Visual comparators are devices for matching the colour of the test sample to the colour standards of a known value. The index number refers to a calibration chart in the instructions for each test.

### 3.3.1 Monitoring parameters

The monitoring parameters are carbon monoxide (CO), hydrogen sulphide ($H_2S$), sulphur dioxide ($SO_2$), nitrogen dioxide ($NO_2$), ammonia ($NH_3$) and chlorine ($Cl_2$).

The basic working principal of this monitoring kit is based on a chemical reaction between specific reagents and individual gaseous pollutants in the reagent solution. These reactions are very specific and require care during the test. In almost all cases, the end of product of each reaction is a coloured clear compound of which the colour intensity is directly proportional to the concentration of the gaseous pollutant being monitored. Time is the most important parameter as it is directly proportional to the amount of gas impinged. Keeping in View this criterion, the monitoring of six gases has been ensured on quantitative bases.

### 3.3.2 The air pump

The air pump has been adopted for air monitoring in the AA-190 version which is a product intended for aquarium use with in domestic environments where air is not plentiful in water. These air pumps are AC/DC operable and have a dual flow of 1.0 and 2.0 l/m, calibrated against the standard flow meter. The AC version is suitable for operations at sites where power supply is available. However, the DC version utilizing a single dry cell, can be carried conveniently with other monitoring devices. Among the important features of the air sampling pump is the impinger tube holder built into the pump housing. Some specifications of these air pumps are as follows.

| | | |
|---|---|---|
| Power | : | AC /DC |
| Power rating | : | 0.2 watts |
| Pumping system | : | Diaphragm type |
| Type of compressor | : | Eccentric piston |
| Operational mode | : | Manual |
| Dimension | : | 12 x 7 x 4 cm$^3$ |
| Construction material | : | High strength plastic |
| Casing | : | Manually locked |
| Air flow rate | : | 1 l/m at 1.5 V |
| | | 2 l/m at 3.0 V |

### 3.1.3 Air pollution monitoring setup

The air pollution monitoring setup consists of tygon tube, glass tube, cork fitted with glass, impinging tubes, impinging tube holder fitted with air pump, switch ON/OFF push button, air pump and manual lock.

### 3.3.4 Plastic and glassware

The plastic and glassware for a one time measurement of six gases are six impinging tubes, six test tubes, six corks fitted with glass tubes, twenty one plastic bottles for chemicals, six syringes, one tygon tube, 6 visual comparators.

### 3.3.5 Composition of reagents for detecting gases

The detection of carbon monoxide is made with two reagents. The first reagent is 4-carboxybenzene sulphonamide (0.8%) mixed with 0.17% of silver nitrate and 2.2% of sodium hydroxide in aqueous solution. The second reagent is 0.1% w/v silver nitrate aqueous solution.

The detection of sulphur dioxide is made with five reagents. The first reagent is a solution of 1.1% mercuric chloride with 0.6% potassium chloride and 0.007% EDTA disodium salt in aqueous solution. The reagent number two is sulphuric acid. The reagent number three is 30% solution of formaldehyde. The reagent number four 0.2% solution of pararosanaline in 1.0 M of hydrochloric acid. Reagent number five is distilled water.

The detection of nitrogen dixide is made with the help of three reagents. The first reagent is a solution of 14% sulphanailic acid with 0.5% glacial acetic acid in aqueous solution. The second reagent is 0.1% N- (1- Naphthyl) - etheline diamine-dihydrochlorate in aqueous solution. The third reagent is 0.5 mg of pure sodium nitrate for each test.

The detection of hydrogen sulphide is made with five reagents. The first reagent is a solution of 4.3% cadmium sulphate with 0.3% sodium hydroxide in aqueous solution. The reagent number two is 63.9 % of sulphuric acid with N, N-dimethylphenylene diamine oxalate or 0.67% dihydrochloride in aqueous solution. The reagent number three is 100% solution of ferric chloride in aqueous solution. The reagent number four is 40% ammonium phosphate in aqueous solution. Reagent number five is distilled water.

The detection of ammonia is made with the help of three reagents. The first reagent is an aqueous solution of 1.0 % sulphuric acid. The second reagent is 15% potassium hydroxide with 3.3% of mercuric chloride and 6.25% of potassium iodide in aqueous solution. The third reagent is distilled water.

The detection of chlorine is made with the help of three reagents. The first reagent is distilled water. The second reagent is 5% w/v aqueous solution of sodium hydroxide. The third reagent is DPD tablets.

## 3.4 Procedure for air pollution analysis:

The unit has been developed for the monitoring of CO, $SO_2$, NO, $H_2S$, $NH_3$ and $Cl_2$ and is described with reference to the monitoring procedures with necessary precautions to be taken during the test. The procedures for various gases are given below.

### 3.4.1 Carbon monoxide

The procedure for detection of carbon monoxide is; 1). pour 10 ml of Carbon monoxide absorbing solution (reagent number one) into an impinging tube. Connect the impinging apparatus to inlet of the sample pump. Make sure that the long tube is immersed into the absorbing solution. Collect the air sample for (10, 20 or 30 minutes) as appropriate. 2). on completion of the air-sampling period, pour the contents of the impinging tube into a clean test tube. Add 4-5 drops of reagent number two. 3). match the colour of the test sample to the index of the colour standard in order to obtain a concentration of carbon monoxide.

The following calibration chart converts the comparator chart index reading into the concentration of carbon monoxide in the atmosphere to parts per million (ppm). The chart is based upon the prescribed sampling period for individual test. When the colour match, find the corresponding index value on the calibration chart. Then read down the line until the sampling time equates.

| Time in minutes | Comparator index number | | | | | | |
|---|---|---|---|---|---|---|---|
| | 1 | 2 | 3 | 4 | 5 | 6 | 7 |
| 10 | 33 | 67 | 100 | 133 | 166 | 200 | 233 |
| 20 | 25 | 50 | 75 | 100 | 125 | 150 | 175 |
| 30 | 20 | 40 | 60 | 80 | 100 | 120 | 140 |

Table 3.1: Carbon monoxide: Calibration chart (ppm)

### 3.4.2 Nitrogen dioxide

The procedure for detection of nitrogen dioxide is; 1). pour 10 ml of reagent number one into the impinging tube. Connect the impinging tube to the air-sampling pump. Make sure that the long tube is immersed in the solution. 2). sample for 1, 5, 10, 15 or 20 minutes as appropriate. At the end of the sampling time pour the contents of the impinging tube into the given test tube. Dilute to 10 ml volume with reagent number one, if any absorbing solution is lost through evaporation during the sample procedure. 4). add 1 drop of reagent number two to the test tube, secure the cap and mix. 5). add one level measure of reagent number three to the content of the test tube. Secure the cap and mix. Wait for 10 minutes for full colour development. 6). check the test samples colour against the index of the colour standard in the comparator. Record the index number, which gives the proper colour match.

The following calibration chart converts the comparator chart index reading into the concentration of nitrogen dioxide in the atmosphere to parts per million (ppm). The chart is based upon the prescribed sampling period for individual tests. When the colour match, find the corresponding index value in the calibration chart. Then read down the line until the sampling time equated.

| Time in minutes | Comparator index number | | | | | | |
|---|---|---|---|---|---|---|---|
| | 1 | 2 | 3 | 4 | 5 | 6 | 7 |
| 1 | 0.00 | 2.80 | 7.00 | 14.0 | 21.00 | 28.00 | 35.0 |
| 5 | 0.00 | 0.56 | 1.40 | 2.80 | 4.20 | 5.60 | 8.40 |
| 10 | 0.00 | 0.28 | 0.70 | 1.40 | 2.10 | 2.80 | 4.20 |
| 15 | 0.00 | 0.19 | 0.47 | 0.93 | 1.40 | 1.87 | 2.80 |
| 20 | 0.00 | 0.14 | 0.35 | 0.70 | 1.05 | 1.40 | 2.10 |

Table 3.2: Nitrogen dioxide: Calibration chart (ppm) [Field instructions manual]

### 3.4.3 Sulphur dioxide

The procedure for detection of sulphur dioxide is; 1). pour 10 ml of reagent number one into the impinging tube (caution: it is highly poisonous reagent). Connect the impinging tube to the air-sampling pump. Make sure that the long tube is immersed in the solution. 2). sample for 10, 30 or 60 minutes as appropriate. Cover the impinging tube in aluminium foil to protect it from light. 3). with the 0.5 g measuring spoon, add one level measure of reagent number two. Mix the content of the tube until the power is dissolved. Allow reacting for 10 minutes. 4). add one drop of reagent number three and mix. Add 5ml (5 measures) of reagent number four. Mix and allow standing for 30 minutes. Dilute to 25 ml with reagent number five. 5). check the test samples colour against the index of the colour standard in the comparator. Record the index number, which gives the proper colour match.

The following calibration chart converts the comparator chart index reading into the concentration of sulphur dioxide in the atmosphere to parts per million (ppm). The chart is based upon the prescribed sampling period for individual tests. When the colour match, find the corresponding index value in the calibration chart. Then read down the line until the sampling time equated.

| Time in minutes | Comparator index number | | | | | | |
|---|---|---|---|---|---|---|---|
| | 1 | 2 | 3 | 4 | 5 | 6 | 7 |
| 10 | 0.00 | 0.30 | 0.90 | 2.10 | 3.90 | 7.80 | 11.70 |
| 30 | 0.00 | 0.20 | 0.30 | 0.70 | 1.30 | 2.60 | 3.90 |
| 60 | 0.00 | 0.10 | 0.20 | 0.40 | 0.70 | 1.30 | 2.00 |

Table 3.3: Sulphur dioxide: Calibration chart (ppm) [Field instructions manual]

### 3.4.4 Hydrogen sulphide

The procedure for detection of hydrogen sulphide is; 1). add 10 ml of reagent number one into the impinging tube connected to the air sampling pump. Make sure that the long tube is immersed in the absorbing solution. 2). collect an air sample for period of 10, 20 or 30 minutes as appropriate. 3). at the end of the sampling period, disconnect the impinging tube from the sampling pump and add 0.5 ml of reagent number two by using 0.5 ml pipette. 4). using the other 0.5 ml pipette, add 0.5 ml of reagent number 3 to the impinging tube and mix. 5). add four drops of reagent number four and mix. A blue colour should be developed. If hydrogen sulphide is present within the detectable limit. 6). after one minute add 1.5ml of reagent number five with the given dropper. 7). transfer the solution to the test tube and compare the colour of sample to the index of the colour match.

The following calibration chart converts the comparator chart index reading into the concentration of hydrogen sulphide in the atmosphere to parts per million (ppm). The chart is based upon the prescribed sampling period for individual tests. When the colour match, find the corresponding index value in the calibration chart. Then read down the line until the sampling time equated.

| Time in minutes | Comparator index number | | | | | | |
|---|---|---|---|---|---|---|---|
| | 1 | 2 | 3 | 4 | 5 | 6 | 7 |
| 10 | 0.60 | 1.40 | 2.80 | 5.50 | 11.10 | 16.60 | 22.20 |
| 20 | 0.30 | 0.70 | 1.40 | 2.08 | 5.50 | 8.30 | 11.10 |
| 30 | 0.20 | 0.50 | 0.90 | 1.90 | 3.70 | 5.50 | 7.40 |

Table 3.4: Hydrogen sulphide: Calibration chart (ppm) [Field instructions manual]

### 3.4.5 Ammonia

The procedure for detection of ammonia is; 1). pour 10 ml of reagent number one into the impinging tube connected to the air sampling pump. Make sure that the long glass tube is immersed in the absorbing solution. 2). collect an air sample for period of 5, 10, 15 or 20 minutes as appropriate. 3). at the end of the sampling period, add the contents of the impinging tube to the 5 ml mark of the given test tube. 4). add two drops of reagent number three to the test tube. Secure the cap and mix for one minute. 5). add eight drops of reagent number three to the test tube. Secure the cap and mix for one minute. The development of yellowish brown colour indicates the presence of ammonia 6). compare the colour of the sample to the index of the colour standards. Record the index number, which gives the proper colour match.

The following calibration chart converts the comparator chart index reading into the concentration of ammonia in the atmosphere to parts per million (ppm). The chart is based upon the prescribed sampling period for individual tests. When the colour match, find the corresponding index value in the calibration chart. Then read down the line until the sampling time equated.

| Time in minutes | Comparator index number | | | | | | |
|---|---|---|---|---|---|---|---|
| | 1 | 2 | 3 | 4 | 5 | 6 | 7 |
| 5 | 10.29 | 2.58 | 30.90 | 41.49 | 51.48 | 61.77 | 72.06 |
| 10 | 5.16 | 10.29 | 15.45 | 20.58 | 25.74 | 30.90 | 36.03 |
| 15 | 3.45 | 6.48 | 10.29 | 13.74 | 17.13 | 20.58 | 24.06 |
| 20 | 2.58 | 5.16 | 7.74 | 10.29 | 12.90 | 15.45 | 18.03 |

Table 3.5: Ammonia: Calibration chart (ppm) [Field instructions manual]

### 3.4.6 Chlorine

The procedure for detection of chlorine is; 1). fill the impinging tube to the 10 ml marked with reagent number one. 2). add two drops of reagent number two to the content of the impinging tube and mix. Connect the impinging tube to the sampling pump. Make sure the long tube is immersed in the absorbing solution. 3). Sample for 15, 30 and 60 minutes as appropriate. 4). at the end of the sampling period, disconnect the impinging tube and pour the contents in the clean test tube. Dilute to the 10 ml mark with reagent number one, in order to replace the liquid loss due to evaporation. Add one reagent tablet number three to the test tube. Secure the cap and shake well until the tablet is resolved. If chlorine is present, a pink to red colour will be developed. Compare the colour of the sample to the index of the colour standards. Record the index number which gives the proper match.

The following calibration chart converts the comparator chart index reading into the concentration of chlorine in the atmosphere to parts per million (ppm). The chart is based upon the prescribed sampling period for individual tests. When the colour match, find the corresponding index value in the calibration chart. Then read down the line until the sampling time equated.

| Time in minutes | Comparator index number | | | | | | |
|---|---|---|---|---|---|---|---|
| | 1 | 2 | 3 | 4 | 5 | 6 | 7 |
| 15 | 0.40 | 0.90 | 1.30 | 1.80 | 2.20 | 3.30 | 4.40 |
| 30 | 0.20 | 0.40 | 0.70 | 0.90 | 1.10 | 1.70 | 2.20 |
| 60 | 0.10 | 0.20 | 0.30 | 0.40 | 0.60 | 0.80 | 1.10 |

Table 3.6: Hydrogen sulphide: Calibration chart (ppm) [Field instructions manual]

## 3.5 Precautions

### 3.5.1 The air pump:

1). Check the battery capacity by the switch to the "ON" position. If the pump makes irregular noise or stalls badly, the battery is probably weak. Replace with new batteries. It is important that the battery should be replaced after 10-15 measurement. When inserting the new batteries, care must be taken to ensure correct polarity.

2). The air pump on the AC must be handled with care. The internal connections must not touched.

### 3.5.2 The monitoring unit:

1). Read all the instructions carefully before start.

2). Read the label on each reagent container prior to use.

3). Avoid contact between the reagent chemical and the human body (use gloves, goggles etc., for protection).

4). Always use the test tube caps or stopper provided, and not fingers, on the test tube during shaking and mixing.

5). Rinse thoroughly the tubes before and after each test.

6). Close tightly all reagent containers immediately after use.

7). Keep all equipment out of reach of children reach.

## 3.6 Troubleshooting

The field instructions manual incorporates several tips for operations of the equipment especially when a malfunction in the kit occurs. Some of the common problems and their possible solutions are as follow.

| Problems | Causes and solutions |
|---|---|
| 1. Air pump does not operate | Faulty switch: dead batteries or motor replace battery. |
| 2. Produces abnormal noise. | Replace battery |
| 3. Airflow efficiency has dropped | Replace diaphragm. |
| 4. Colour developed in the impinging reaction either very weak or faint. | Reagents outdate; use fresh reagents |
| 5. Irregular colour development | Reagents contaminated, use fresh reagents. |
| 6. Colour too fast | High concentration of gas, use proper dilution |
| 7. No colour development | Pollutant below detection limit; increase impinging time. |

Table 3.7: Common operational problems and their solutions [Field instructions manual]

## 3.7 Visual comparators

There are devices that are used for the comparison of colour strength developed as a result of the chemical reaction of reagents used.

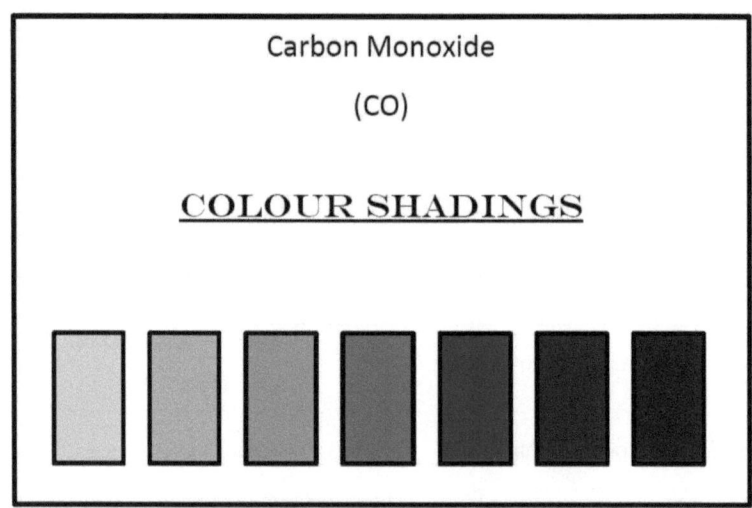

Figure 3.1: The visual comparator for carbon monoxide (Field instructions manual)

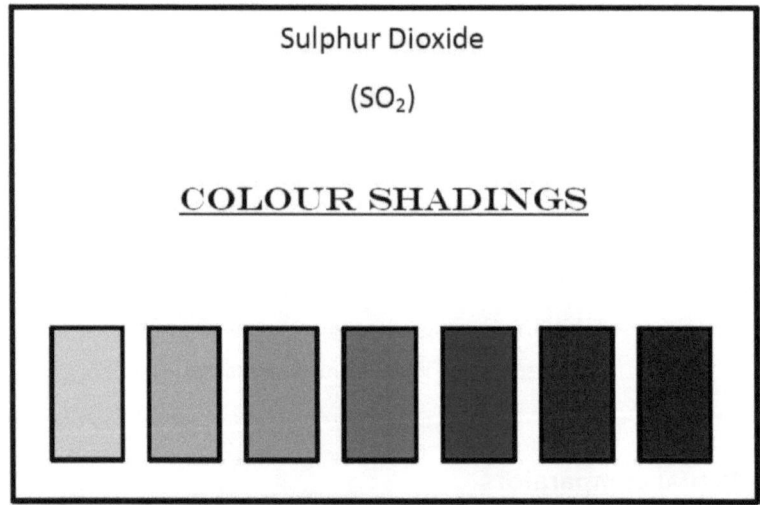

Figure 3.2: The visual comparator for sulphur dioxide (Field instructions manual)

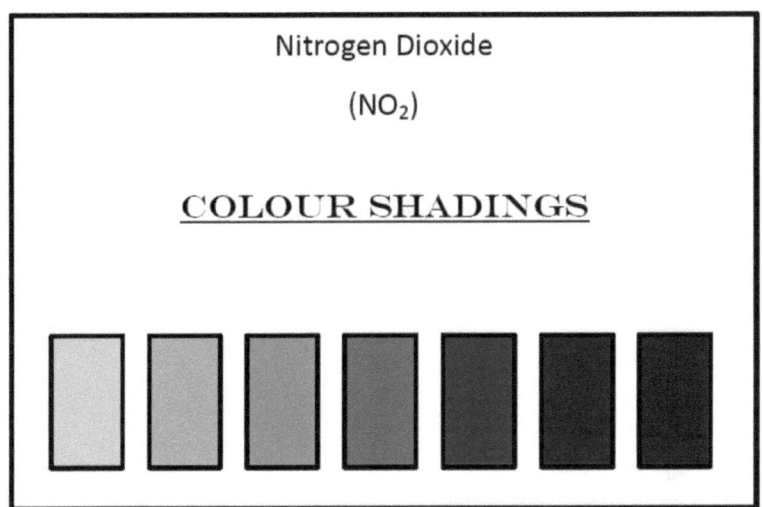

Figure 3.3: The visual comparator for nitrogen dioxide (Field instructions manual)

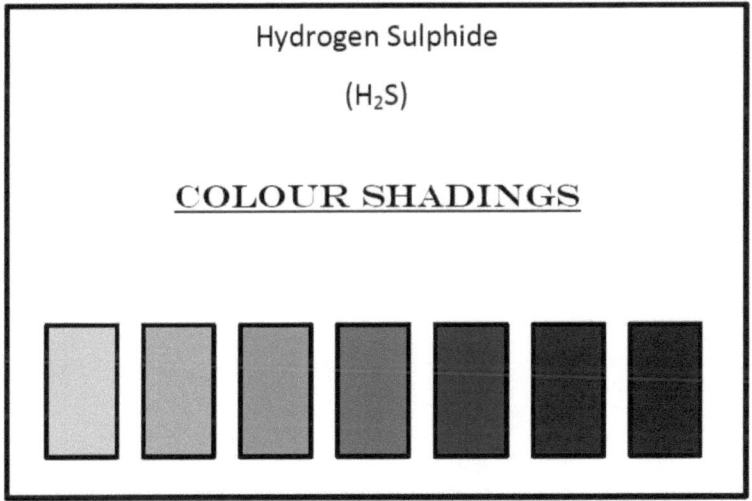

Figure 3.4: The visual comparator for hydrogen sulphide (Field instructions manual)

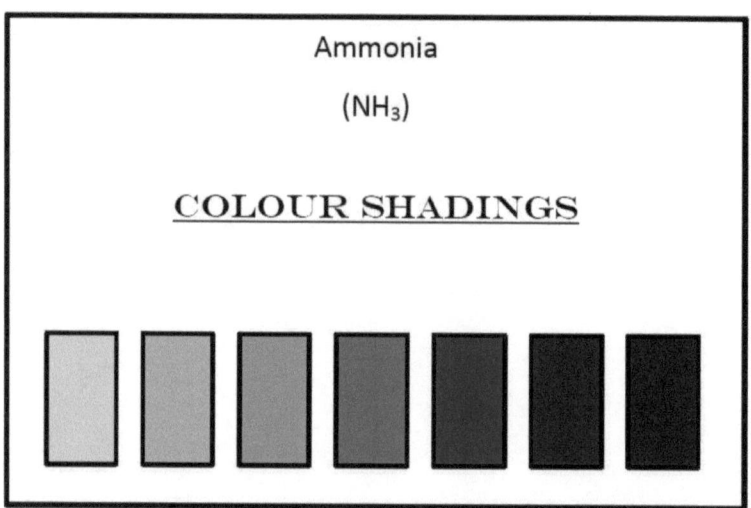

Figure 3.5: The visual comparator for ammonia (Field instructions manual)

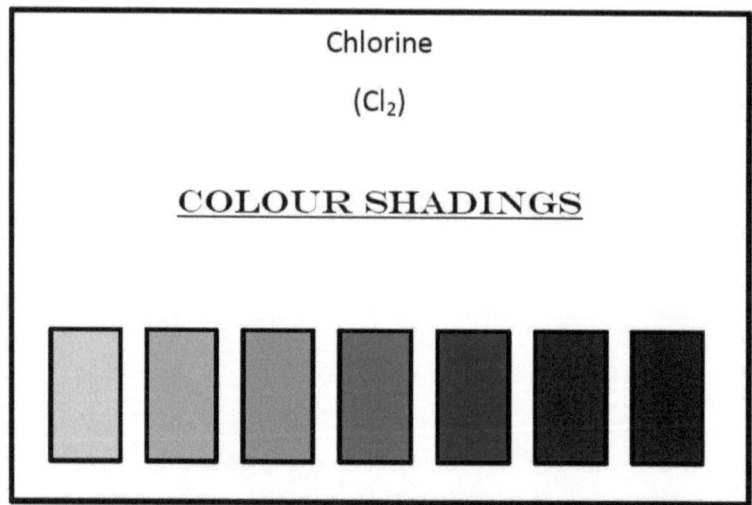

Figure 3.6: The visual comparator for chlorine (Field instructions manual)

# Chapter 4:
# Air pollution sampling and analysis

## 4.1 Importance of monitoring air pollution

Monitoring provides raw measurements of air pollutant concentrations. With appropriate analysis and interpretation, these measurements can be transformed into useful information on the quality of the air. This information has many uses, the principal ones are listed below.

Air quality measurements can provide essential data to help answer a wide variety of questions such as: How bad is air pollution? How does air pollution varies with time and location? Is air pollution getting better or worse?

Many people need to know how bad air pollution is in their area. The general people, worried about pollution or with health problems worsened by air pollution, can benefit from up-to-date information about air quality. Measurements made at monitoring sites across the country can be interpreted ad relayed to the public via a variety of media outlets such as television, tele-text, telephone helpline or the internet.

Government policy on the environment can only be determined in the light of sufficient information. Knowledge of air quality, and how it varies with time and location from measurements can help governments to decide on policies to decrease pollution levels.

## 4.2 Importance of data acquisition and analysis

The air pollution data acquisition is a pre-requisite for the environmental analysis of particular area and finding the exposure extent of that area to health. The monitoring component of the study is concentrated on the collection of the air pollutant data by using the techniques of absorption of these gases in different chemical reagents whereas the comparison has been made with the data monitored in the past year. Data for both years has been compared against the standard values given by WHO. The areas we have monitored are exposed to air pollution to great extent. Monitoring of air pollution has been performed. Pollutant data is obtained and analysed for different areas in the city of Lahore. The city is seriously exposed to environmental dangers by different air pollutants due to high population and hence quality of life has been affected.

The evolution of life has always been associated with the evaluation of atmosphere. The characteristics and composition of our air depends on the pattern of life that forms on the earth and depends upon the interacting system. The interaction of human and also the other millions of species interacting with the atmosphere describes this interacting system.

Emissions of highly toxic substances are given special consideration. Although some of these pollutants are also produced by nature, the main environmental problem result from human activities in residential and industrial areas. Air pollutants are often transported over considerable distances, affecting air quality, ecosystems, lakes and other surface water, groundwater, soils and buildings in adjacent and distant countries.

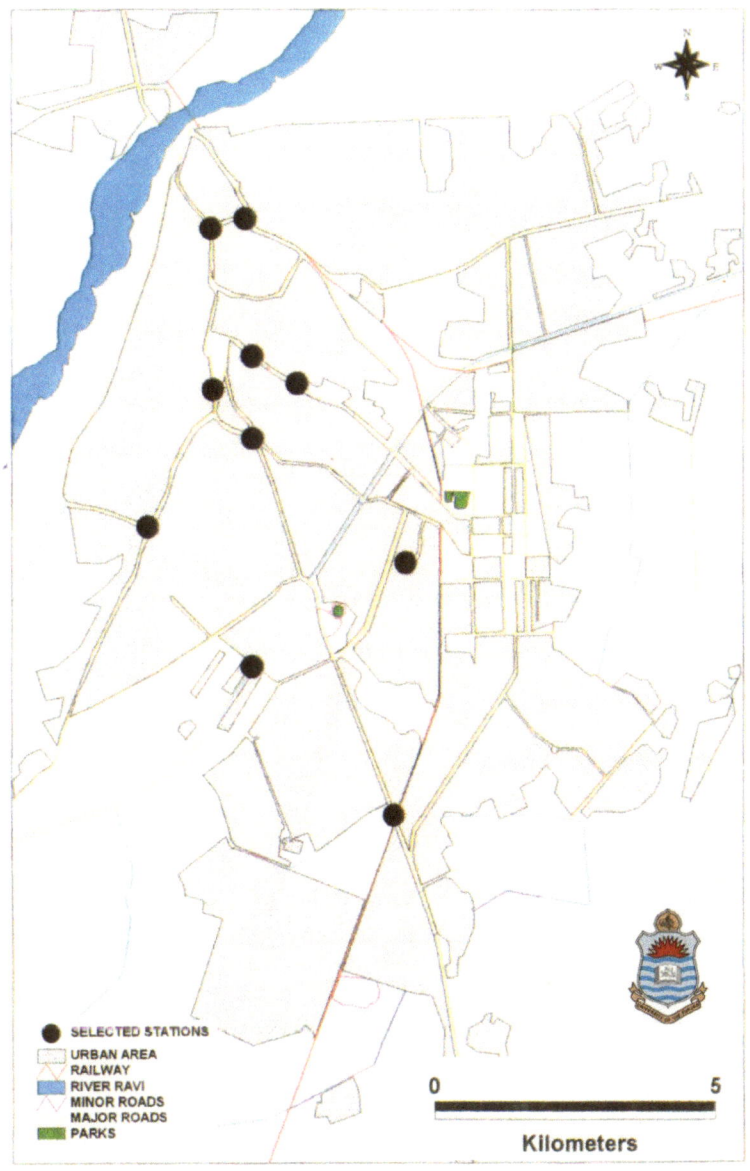

Figure 4.1: Map of Lahore city showing sampling sites loaded into a GIS program (Isma 2000).

The air pollutant data has been collected at the following locations of the Lahore city.

1. Badami bagh
2. Ichhra market
3. Main market "Gulberg"
4. Railway station
5. Shadman market

These areas are estimated to be exposed to air pollution to a great extent. Monitoring of air pollution has been performed and data has been obtained and analysed for the above mentioned sites.

## 4.3 Exposure level study procedure

The exposure level studies can be classified into three groups:

### 4.3.1 Human exposure studies

Human exposure studies are based on the occupational or accidental exposures resulting in adverse effects. The studies also include subjects exposed to atmospheric pollutants in uncontrolled conditions. For example the subjects include truck drivers, traffic policemen and individuals working in toll-booths are used to determine cause and effect relationships. Data collected from major air pollution episodes are used to establish a relationship between air pollution and reported diseases.

### 4.3.2 Laboratory research / animal studies

This is conducted on animals and cells or biological systems. Actual exposure experiment are carried out. Different types of animals are critically studied under controlled conditions of concentrations and dosages. This provides immense information regarding the mode of various pollutants and their effects. Statistical analysis is rigorously used. The problem, however is that the analysis of these results to the human population is not precise because of high number of unknown sources. Moreover, one can only test one cause and effect one relationship at a time.

### 4.3.3 Epidemiological studies

In epidemiological studies, the relationship between the distribution of specific diseases in a human population and possible causes are determined. The studies focus on communities. A typical study uses hospital records and abnormality records. To avoid misinterpretation, the population under study must be carefully observed for smoking habits, occupational exposures and any other factor that might prejudice the results of the study. The studies are not repeatable and often have no control over timing and other variables. A typical epidemiological study involves the collection, compilation and analysis of data.

## 4.4  WHO concentration thresholds

The major purpose of the field work is to measure the concentration and to compare it with the past year data, whether it is higher or lower than the past data. The maximum concentration allowance, which is assigned by WHO, is as given

| Name of gas | Conc. (ppm) | Effects on human health |
|---|---|---|
| Carbon monoxide | 50 | Headache and de-oxygenation of blood |
| Sulphur dioxide | 2 | Bronchitis and shortness of breath |
| Nitrogen dioxide | 3 | Acute tract infection and increases mortality |
| Hydrogen sulphide | 10 | Elimination of enzymes based metals |
| Ammonia | 25 | Respiratory tissue damage |
| Chlorine | 1 | Choking and irritation of respiratory system |

Table 4.1: Allowable limits of air pollutants

## 4.5 Atmospheric concentration units

There are two concentration units that are commonly used in reporting atmospheric species, viz., 1). microgram per cubic meter, 2). parts per million.

The ppm unit can be further sub-divided into

1). Parts per hundred millions by volume (pphm)

2). Parts per billion by volume (ppb)

3). Parts per trillion (ppt)

## 4.6 Air pollution at Badami Bagh

### 4.6.1 Map of the sampling site

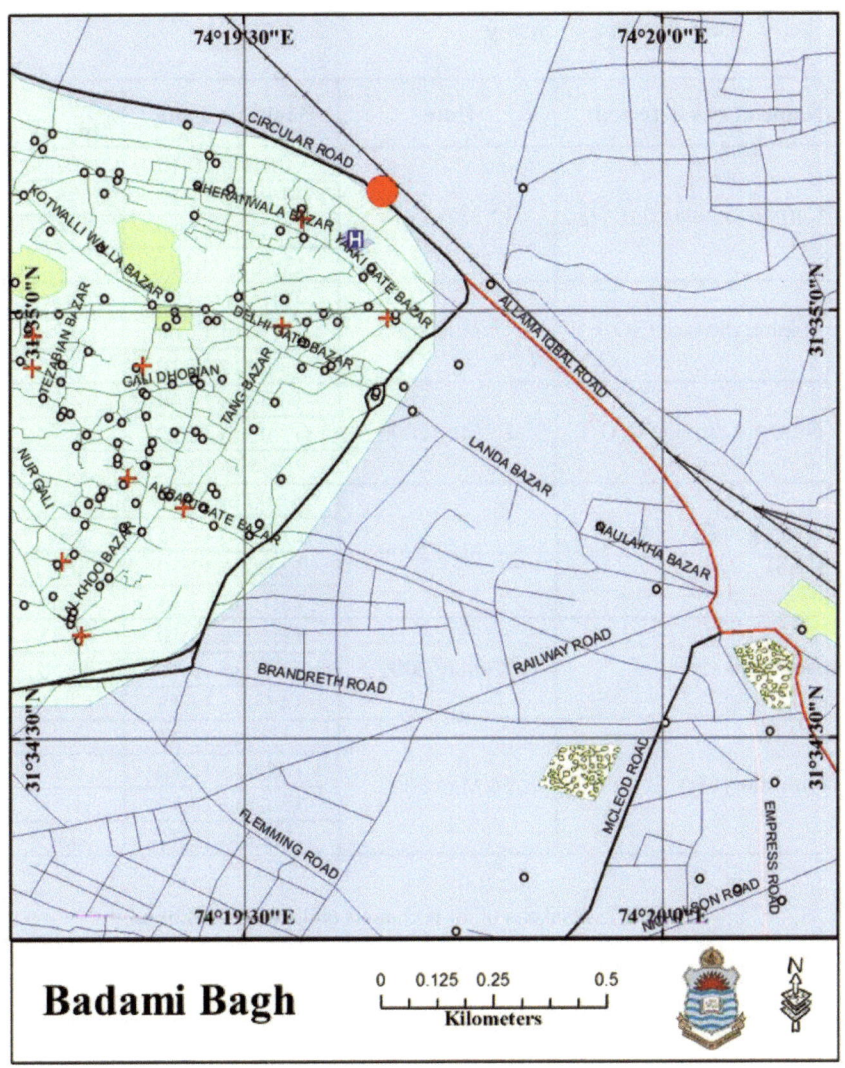

Figure 4.2: Geographical Map of Badami Bagh

## 4.6.2 Data acquisition and plotting

Location: Badami Bagh
Day conditions: Sunny

| Name of gas detected | Date | Monitor Time | Conc. (ppm) |
|---|---|---|---|
| Carbon monoxide (CO) | 15 May 2000 | 10:00 to 10:10 | 67 |
| | | 11:00 to 11:20 | 75 |
| | | 12:30 to 12:50 | 75 |
| | | 13:40 to 14:10 | 80 |
| Sulphur dioxide ($SO_2$) | 15 May 2000 | 10:15 to 10:25 | 0.3 |
| | | 11:30 to 12:00 | 0.7 |
| | | 13:00 to 13:10 | 0.9 |
| Nitrogen dioxide ($NO_2$) | 15 May 2000 | 10:30 to 10:40 | 2.1 |
| | | 12:05 to 12:20 | 1.4 |
| | | 13:15 to 13:30 | 1.87 |
| Hydrogen sulphide ($H_2S$) | 23 May 2000 | 11:00 to 11:10 | Nil |
| | | 11:15 to 11:35 | Nil |
| | | 12:00 to 12:30 | 0.9 |
| | | 13:15 to 13:15 | 0.5 |
| Ammonia ($NH_2$) | 23 May 2000 | 11:40 to 11:50 | 15.45 |
| | | 13:00 to 13:10 | 20.58 |
| | | 14:00 to 14:15 | 17.13 |
| Chlorine ($Cl_2$) | 24 May 2000 | 11:30 to 11:45 | Nil |
| | | 12:00 to 12:30 | 0.7 |
| | | 13:00 to 13:30 | 1.1 |
| | | 14:15 to 14:45 | 0.9 |

Table 4.2: Analysis of air pollutants and data acquisition

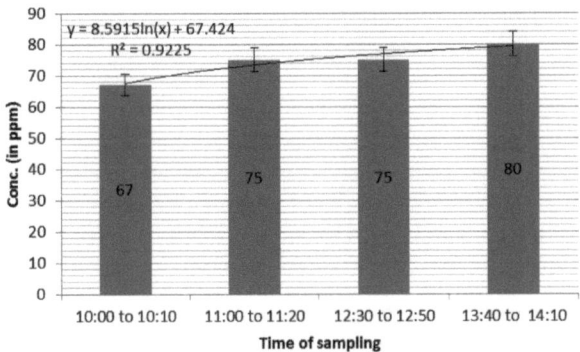

Figure 4.2.1: Carbon monoxide at Badami Bagh

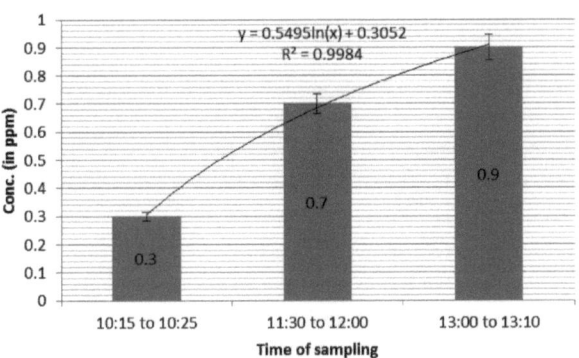

Figure 4.2.2: Sulphur dioxide at Badami Bagh

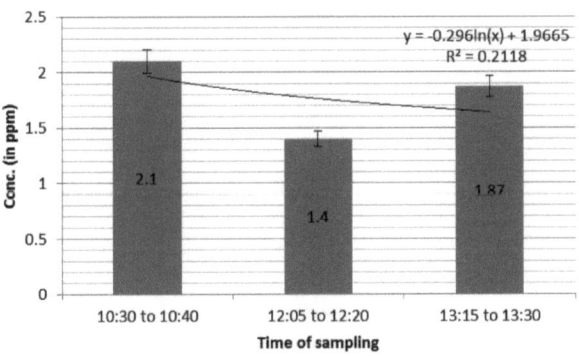

Figure 4.2.3: Nitrogen dioxide at Badami Bagh

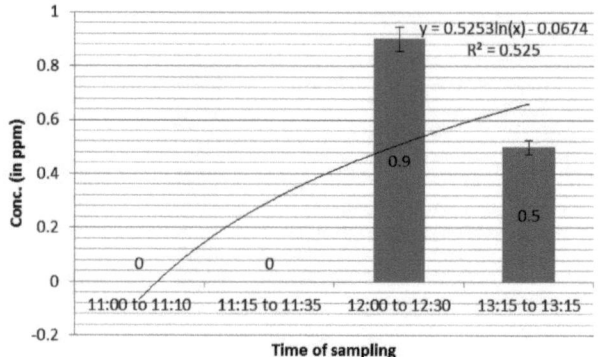

Figure 4.2.4: Hydrogen sulphide at Badami Bagh

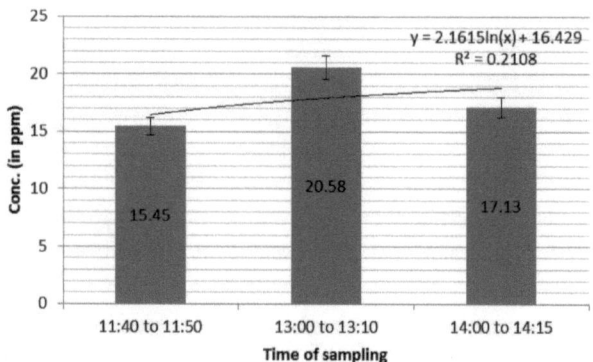

Figure 4.2.5: Ammonia at Badami Bagh

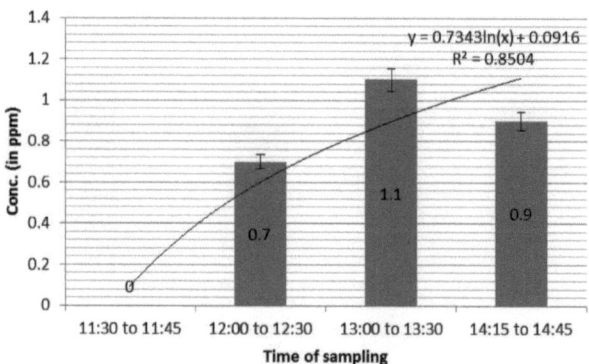

Figure 4.2.6: Chlorine at Badami Bagh

### 4.6.3 Interpretation and trend analysis

The concentration of CO, $SO_2$, $NO_2$, $H_2S$, $NH_3$ and $Cl_2$ has been measured as shown in graphs.

In graph 4.2.1, the concentration of carbon monoxide shows an increasing trend at the Badami Bagh. We can see from the graph that all the values are higher than the WHO standard. The reason for this higher trend may be that the Badami Bagh is very narrow place and the traffic flow is much higher and also the use of fuel in the furnace of iron near in the market increases the concentration of carbon monoxide.

Sulphur dioxide shows a sharp increasing trend as shown in the graph 4.2.2; this may be also due to the increase of traffic density, human and industrial activities and burning of wastes near the chowk.

The concentration of nitrogen dioxide in the morning shows an increasing trend in graph 4.2.3. This may be due to the increase of traffic density with the time producing nitrogen dioxide in large quantities. As the Badami Bagh is a congested area and blocking of traffic causes to accumulate the gas into the air. In afternoon it shown a little decrease, this may be due to the reaction with sunlight.

The concentration of hydrogen sulphide was found after taking several observations. This may be due to the reason that in the morning the presence of gas is lower and in the afternoon the gas shows a higher concentration which may be due to the presence of some marshy places and burning of waste or plastic producing hydrogen sulphide as shown in graph 4.2.4.

In graph 4.2.5, ammonia shows a normal increasing trend in the graph, this may be due to the human activities like to that of use of ammonia as cleansing agent and also by the automobile exhaust which also contribute in the production of ammonia.

In graph 4.2.6, chlorine also shows the positive value. This may be due to the burning of waste and some human activities that produces aerosols such as CFCs.

## 4.7 Air pollution at Ichhra Market

### 4.7.1 Map of the sampling site

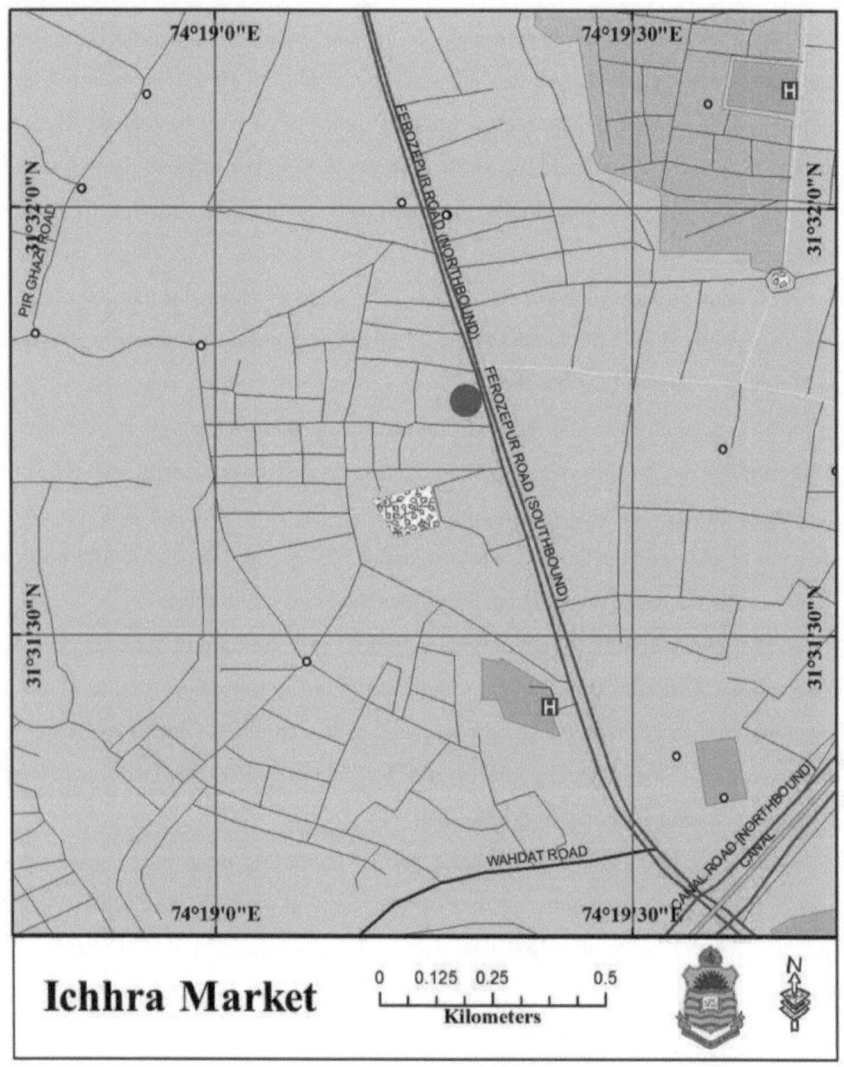

Figure 4.3: Geographical Map of Ichhra Market

## 4.7.2 Data acquisition and plotting

Location: Ichhra Market
Day conditions: Sunny

| Name of gas detected | Date | Monitor Time | Conc. (ppm) |
|---|---|---|---|
| Carbon monoxide (CO) | 12 May 2000 | 10:00 to 10:10 | 67 |
| | | 11:10 to 11:30 | 62 |
| | | 12:35 to 12:55 | 75 |
| Sulphur dioxide (SO2) | 12 May 2000 | 10:15 to 10:45 | 0.7 |
| | | 11:35 to 12:05 | 1.3 |
| | | 13:00 to 13:10 | 0.9 |
| Nitrogen dioxide ($NO_2$) | 12 May 2000 | 10:50 to 11:05 | 0.47 |
| | | 12:15 to 12:30 | 0.93 |
| | | 13:15 to 13:35 | 0.7 |
| Hydrogen Sulphide ($H_2S$) | 26 May 2000 | 9:30 to 9:40 | Nil |
| | | 9.40 to 10:00 | Nil |
| | | 11:00 to 11:30 | Nil |
| | | 13:00 to 13:30 | Nil |
| Ammonia ($NH_3$) | 26 May 2000 | 10:15 to 10:25 | 20.58 |
| | | 11:40 to 11:50 | 15.45 |
| | | 12:40 to 12:55 | 13.75 |
| | | 13:45 to 14:00 | 17.17 |
| Chlorine ($Cl_2$) | 26 May 2000 | 10:30 to 10:45 | Nil |
| | | 12:00 to 12:30 | Nil |
| | | 14:30 to 15:00 | Nil |

Table 4.3: Analysis of air pollutants and data acquisition

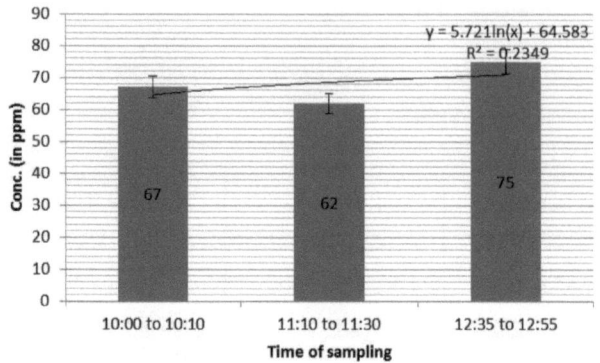

Figure 4.3.1: Carbon monoxide at Ichhra Market

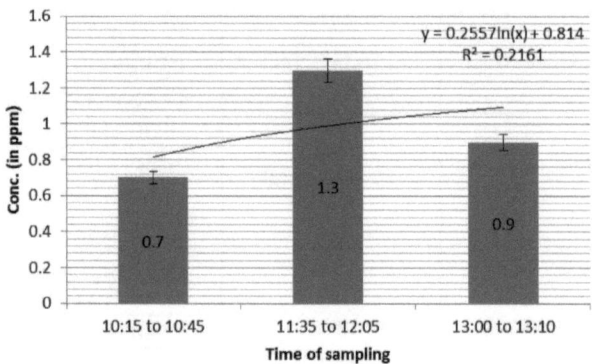

Figure 4.3.2: Sulphur dioxide at Ichhra Market

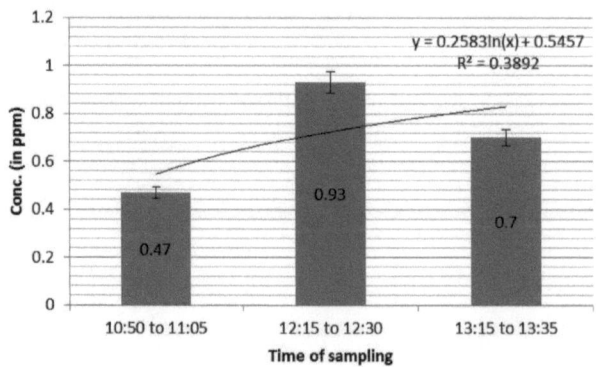

Figure 4.3.3: Nitrogen dioxide at Ichhra Market

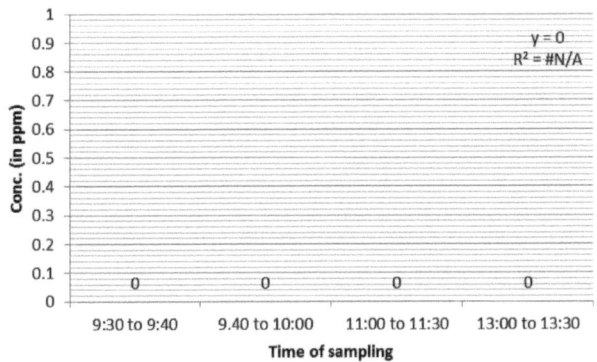

Figure 4.3.4: Hydrogen sulphide at Ichhra Market

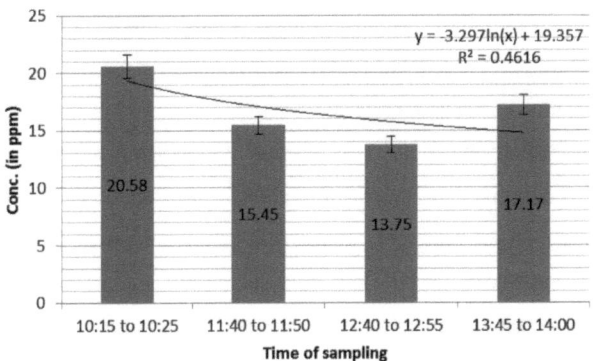

Figure 4.3.5: Ammonia at Ichhra Market

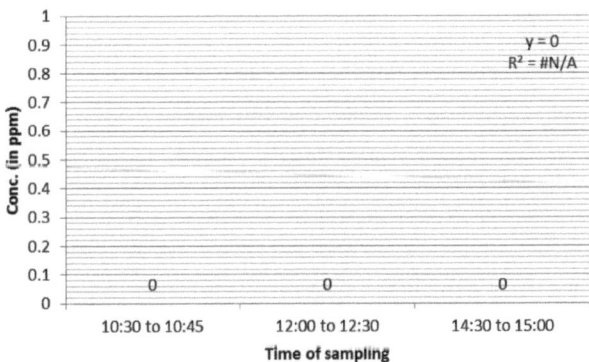

Figure 4.3.6: Chlorine at Ichhra Market

### 4.7.3 Interpretation and trend analysis

The concentration of CO, $SO_2$, $NO_2$, $H_2S$, $NH_3$ and $Cl_2$ has been measured at Ichhra Market. Of these CO, $SO_2$, $NO_2$ and $NH_3$ are found as shown in graphs.

The concentration of carbon monoxide in the graph 4.3.1, shows an increasing trend because Ichhra is an area where the traffic flow is much higher and the carbon monoxide from the vehicle exhaust increase the concentration of the gas which may be the reason for the concentration of carbon monoxide.

As shown in the graph 4.3.2, the concentration of sulphur dioxide shows a increasing trend in the morning but a decreasing trend in the afternoon. The reason for this may be due to increase in human activities with the passage of time. The burning of fuel and natural gas increase the concentration of the sulphur dioxide higher. The slight decreasing trend may be due to some natural chemical reactions such as the reaction of the gas with sunlight.

The concentration of nitrogen dioxide also shows an increasing trend in morning and decreasing trend in the afternoon. It may be due to increase of number of vehicles in the on the road and human activities like the use of natural gas and oil for cooking in the market. It shows a little decrease in the afternoon which may be due to decrease of these activities and traffic (in comparison with morning measurements). This is shown in graph 4.3.3.

No colour is developed in sampling hydrogen sulphide. This may be due to two reasons. Firstly the reagent composition is not right but the reagent was prepared a number of times. Secondly, this may be due to the absence (or presence in non-detectable range) of hydrogen sulphide at this location. After taking a number of observations the concentration of hydrogen sulphide is not found. It means that concentration of hydrogen sulphide is so small not to lie in the detectable range.

In graph 4.3.5, ammonia shows a presence at Ichhra. It shows higher trend in the morning and lower trend in the afternoon. This may be due to more traffic in the morning because at Ichhra market there is a nearby bus stop.

As shown in graph 4.3.6, no colour showing absence of chlorine appeared.

## 4.8 Air pollution at Main Market Gulberg

### 4.8.1 Map of the sampling site

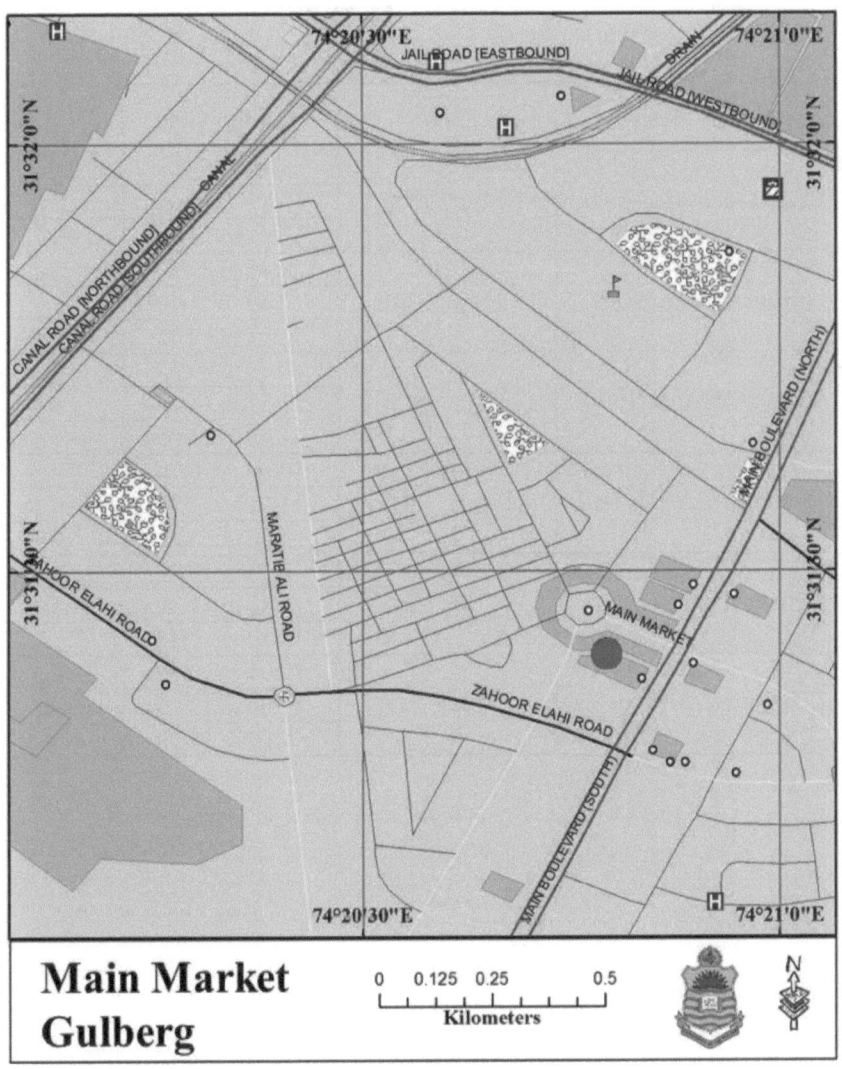

Figure 4.4: Geographical Map of Main Market Gulberg

## 4.8.2 Data acquisition and plotting

Location: Main Market Gulberg
Day conditions: Sunny

| Name of gas detected | Date | Monitor Time | Conc. (ppm) |
|---|---|---|---|
| Carbon monoxide (CO) | 5 May 2000 | 9:30 to 9:40 | 33 |
| | | 10:15 to 10:35 | 50 |
| | | 11:30 to 12:00 | 40 |
| Sulphur dioxide (SO2) | 5 May 2000 | 9:45 to 9:55 | 0.3 |
| | | 10:40 to 11:10 | 0.2 |
| | | 12:05 to 12:35 | 0.3 |
| Nitrogen dioxide ($NO_2$) | 5 May 2000 | 10:00 to 10:10 | 0.3 |
| | | 11:15 to 11:30 | 0.45 |
| | | 12:45 to 13:05 | 0.7 |
| Hydrogen Sulphide ($H_2S$) | 17 May 2000 | 10:00 to 10:10 | Nil |
| | | 11.00 to 11:20 | Nil |
| | | 13:00 to 13:20 | Nil |
| Ammonia ($NH_3$) | 17 May 2000 | 10:15 to 10:25 | 15.45 |
| | | 11:30 to 11:40 | 20.58 |
| | | 13:30 to 13:45 | 17.13 |
| Chlorine ($Cl_2$) | 17 May 2000 | 10:35 to 10:50 | Nil |
| | | 12:00 to 12:15 | Nil |
| | | 14:00 to 14:30 | Nil |

Table 4.4: Analysis of air pollutants and data acquisition

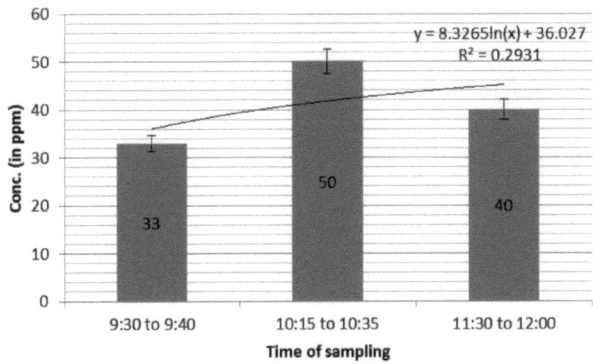

Figure 4.4.1: Carbon monoxide at Main Market Gulberg

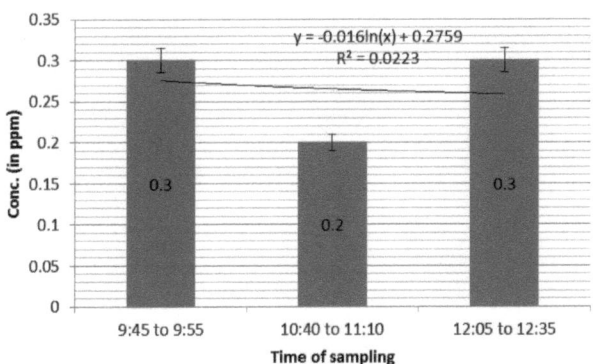

Figure 4.4.2: Sulphur dioxide at Main Market Gulberg

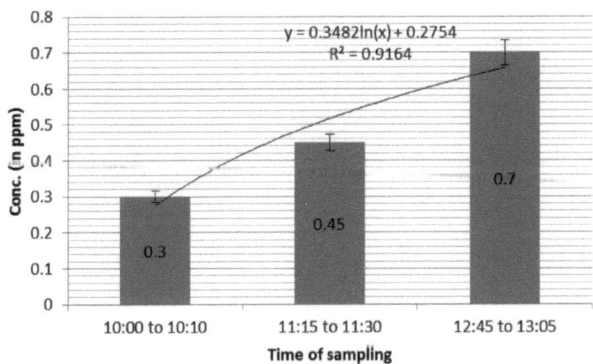

Figure 4.4.3: Nitrogen dioxide at Main Market Gulberg

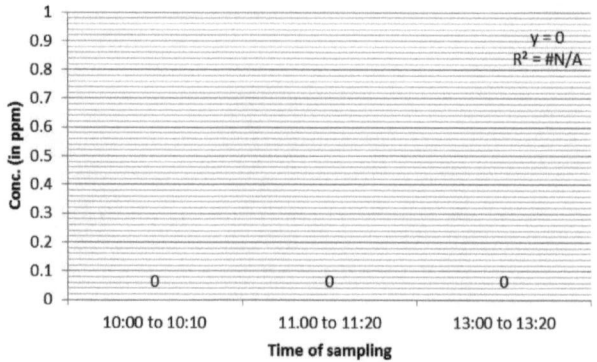

Figure 4.4.4: Hydrogen sulphide at Main Market Gulberg

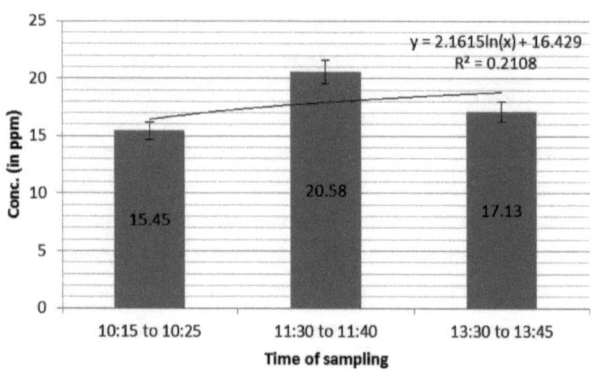

Figure 4.4.5: Ammonia at Main Market Gulberg

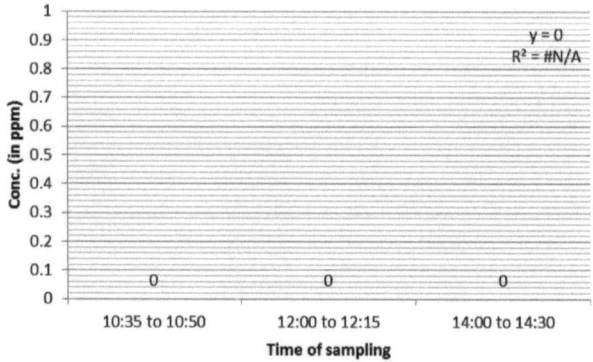

Figure 4.4.6: Chlorine at Main Market Gulberg

### 4.8.3 Interpretation and trend analysis

The concentration of CO, $SO_2$, $NO_2$, $H_2S$, $NH_3$ and $Cl_2$ has been measured at Main Market Gulberg as shown in graphs.

In Main Market the concentration of carbon monoxide and sulphur dioxide shows an increasing trend. The sound reason may be that, at the beginning of the day, the number of vehicles increases gradually and exhaust collected in the atmosphere raise CO and $SO_2$ levels.

The concentration of nitrogen dioxide in the morning shows an increasing trend with time. This may be due to the increase in number of vehicles and human activities such as cooking of foods and other things by the use of natural gas and oil which also contribute to indoor air pollution as shown in the graph 4.4.3.

As shown in the graph 4.4.4, the concentration of hydrogen sulphide at Main Market Gulberg shows a nil value. This may be due to the following reasons; The Gulberg is a commercial market as well as well-planned residential area. Moreover, there is not any garbage dumping place and liquid waste sludge.

In Main Market Gulberg the concentration of ammonia shows an increasing trend in the morning however the exposure level is within the permitted range and all these value are not much harmful to humans.

As shown in the graph 4.4.6, no colour was developed in sampling chlorine. This may be due to two reasons; firstly the reagent formation could not be correct but the reagents were prepared a number of times, secondly, this may be due to the absence and presence in non-detectable range of chlorine at the location. After taking a number of observations the concentration of chlorine has not been detected so it may be concluded that there is no permanent source of chlorine in Main Market Gulberg.

## 4.9 Air pollution at Railway Station

### 4.9.1 Map of the sampling site

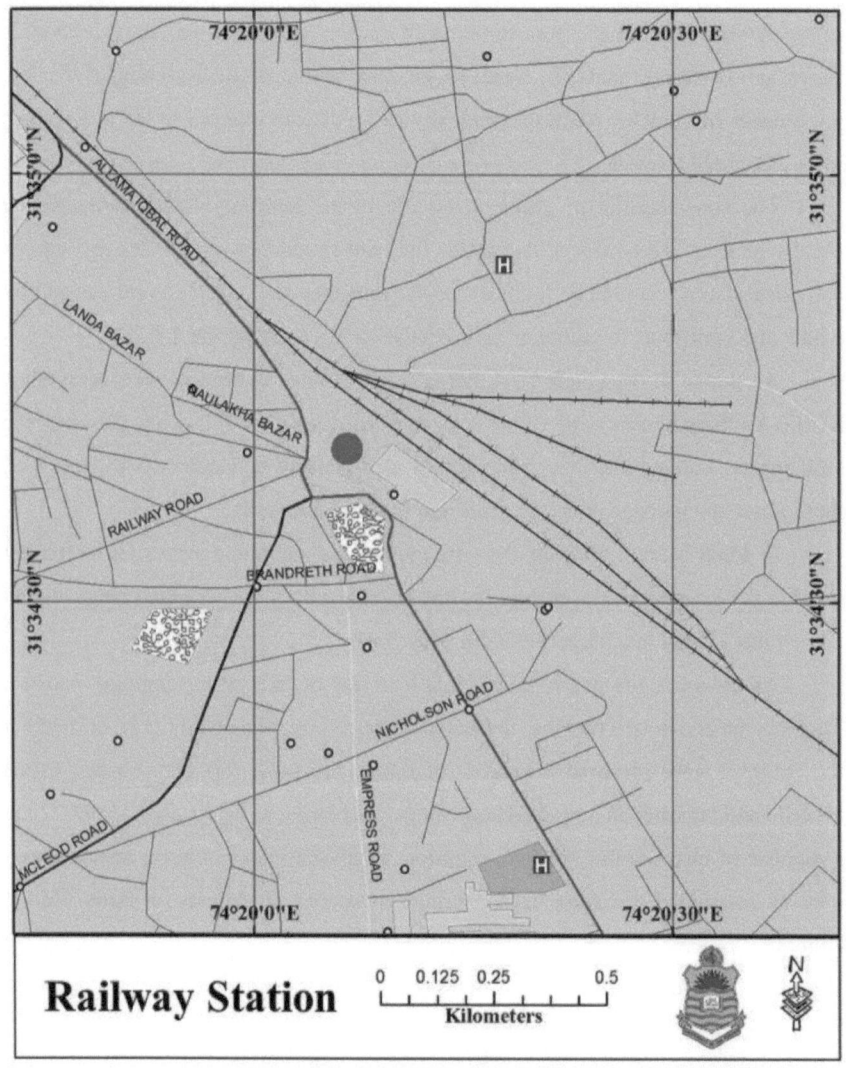

Figure 4.5: Geographical Map of Railway Station

## 4.9.2 Data acquisition and plotting

Location: Railway Station
Day conditions: Sunny

| Name of gas detected | Date | Monitor Time | Conc. (ppm) |
|---|---|---|---|
| Carbon monoxide (CO) | 8 May 2000 | 10:00 to 10:20 | 50 |
| | | 12:00 to 12:30 | 60 |
| | | 14:00 to 14:20 | 75 |
| Sulphur dioxide ($SO_2$) | 8 May 2000 | 10:15 to 10:25 | 2.1 |
| | | 12:45 to 13:15 | 1.3 |
| | | 14:30 to 15:00 | 2.6 |
| Nitrogen dioxide ($NO_2$) | 9 May 2000 | 10:00 to 10:15 | 1.87 |
| | | 11:00 to 11:10 | 2.45 |
| | | 12:30 to 12:40 | 1.75 |
| | | 14:00 to 14:10 | 2.1 |
| Hydrogen Sulphide ($H_2S$) | 20 May 2000 | 10:00 to 10:10 | Nil |
| | | 11:00 to 11:20 | Nil |
| | | 13:00 to 13:30 | 0.5 |
| Ammonia ($NH_3$) | 20 May 2000 | 10:20 to 10:30 | 20.58 |
| | | 11:30 to 11:40 | 25.74 |
| | | 12:30 to 12:45 | 17.13 |
| | | 13:40 to 13:55 | 20.58 |
| Chlorine ($Cl_2$) | 20 May 2000 | 10:40 to 10:55 | 0.4 |
| | | 11:45 to 12:00 | 0.9 |
| | | 14:00 to 14:30 | 0.7 |

Table 4.5: Analysis of air pollutants and data acquisition

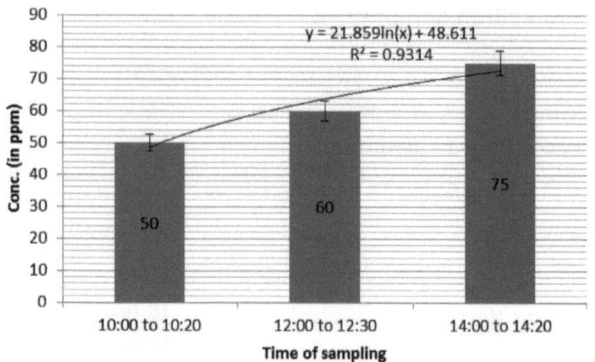

Figure 4.5.1: Carbon monoxide at Railway Station

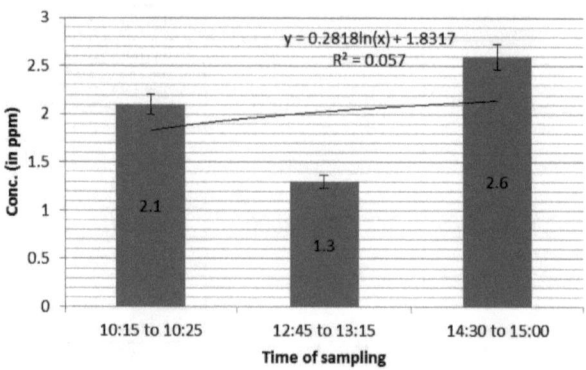

Figure 4.5.2: Sulphur dioxide at Railway Station

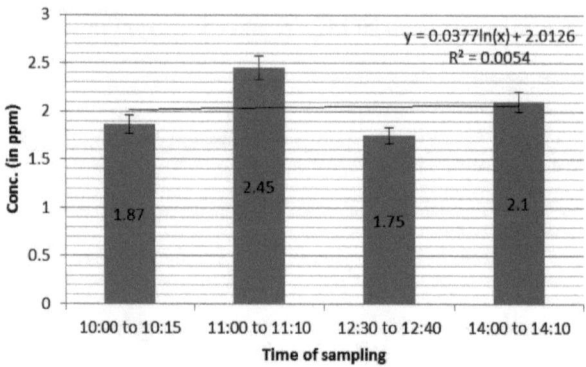

Figure 4.5.3: Nitrogen dioxide at Railway Station

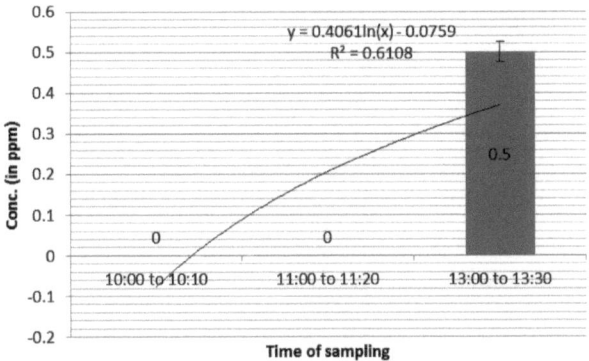

Figure 4.5.4: Hydrogen sulphide at Railway Station

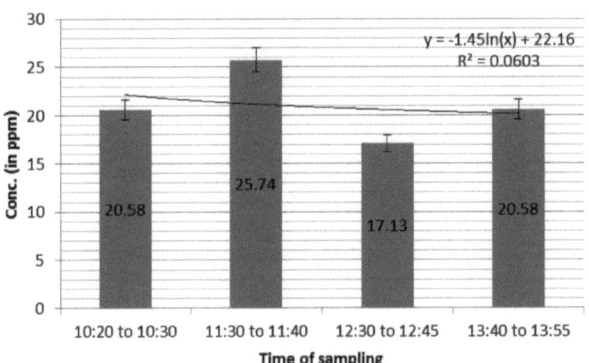

Figure 4.5.5: Ammonia at Railway Station

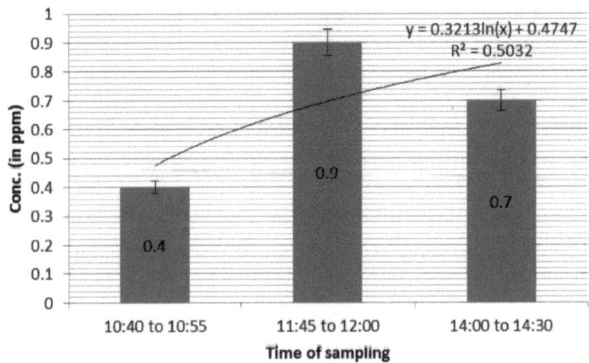

Figure 4.5.6: Chlorine at Railway Station

### 4.9.3 Interpretation and trend analysis

The railway station of Lahore is a commercial area of Lahore city. The concentration of CO, $SO_2$, $NO_2$, $H_2S$, $NH_3$ and $Cl_2$ has been measured at railway station as shown in graphs.

As shown in graph number 4.5.1, the concentration of carbon monoxide shows a sharp increasing trend with time. As the time passes the number of vehicle increases because railway station is a busiest area. At railway station the smoke of car, buses, wagons and particularly the train locomotives produce such a great quantity of carbon monoxide in the atmosphere.

Sulphur dioxide also shows a sharp decreasing trend. From the graph 4.5.2, at 11 am, it reaches its peak value. It is due to increase of vehicle density that uses fuel, which has sulphur contents, and also due to the use of oil and natural gas those produce sulphur dioxide in the atmosphere.

The concentration of $NO_2$ in the morning shows an increasing trend as shown in figure 4.5.3, but a little decrease in the afternoon; this may be due to some atmospheric chemical reaction in which the reaction of gas with sunlight.

The concentration of hydrogen sulphide shows an increasing trend in the afternoon. This may be due to the burning of waste and some other materials or due to the gases emitted from the solid waste present in the area.

Ammonia also shows an increasing trend. This may be due to the use of ammonia as cleansing agent or naturally breaking down of organic matter which is present due to the animals in the area used in the carts etc.

As shown in graph number 4.5.6, chlorine also show an increasing trend in the morning, this may be due to the use of chlorine containing compounds such as CFC's and aerosol propellant. But a little decrease in the afternoon because human activities that are responsible for the production of aerosols decrease in the afternoon.

## 4.10 Air pollution at Shadman Market

### 4.10.1  Map of the sampling site

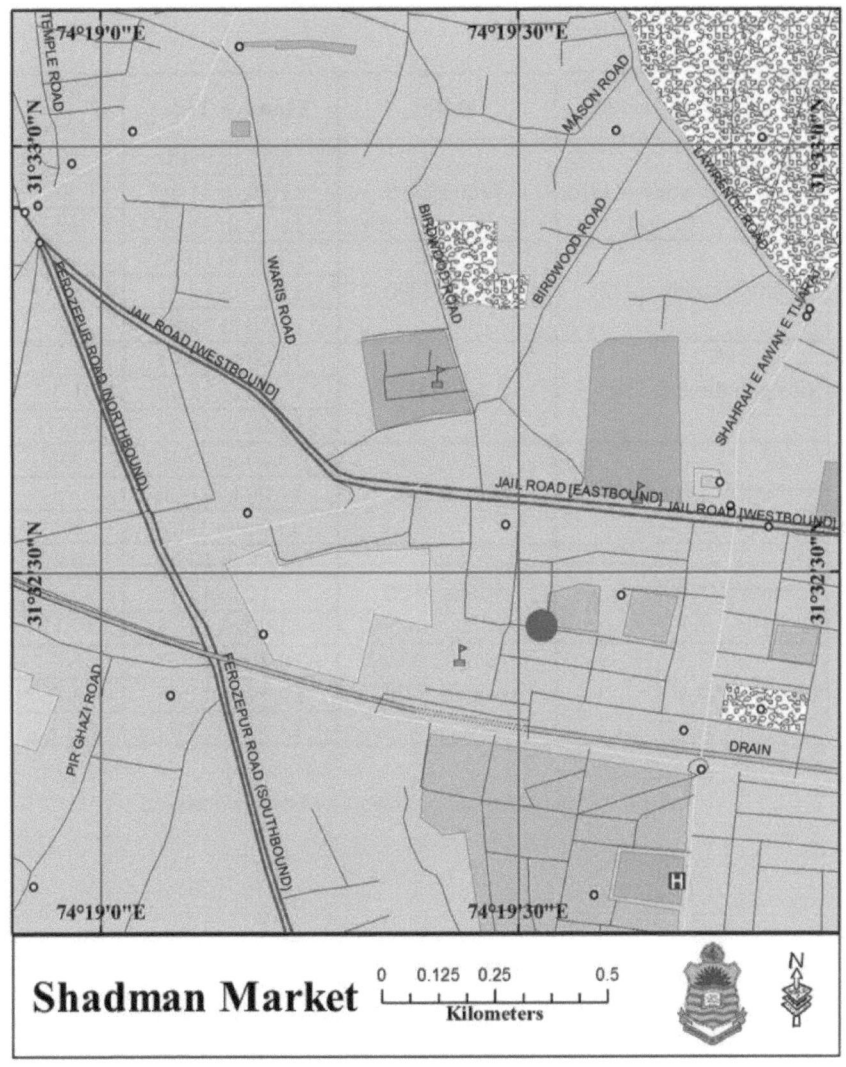

Figure 4.6: Geographical Map of Shadman Market

## 4.10.2 Data acquisition and plotting

Location:       Shadman Market
Day conditions: Sunny

| Name of gas detected | Date | Monitor Time | Conc. (ppm) |
|---|---|---|---|
| Carbon monoxide (CO) | 3 June 2000 | 9:30 to 9:40 | 67 |
|  |  | 11:00 to 11:20 | 50 |
|  |  | 13:30 to 14:00 | 60 |
| Sulphur dioxide (SO2) | 3 June 2000 | 10:00 to 10:30 | 0.2 |
|  |  | 11:15 to 12:15 | 0.1 |
|  |  | 14:30 to 15:00 | 0.3 |
| Nitrogen dioxide ($NO_2$) | 2 June 2000 | 10:00 to 10:15 | 1.4 |
|  |  | 12:00 to 12:10 | 1.4 |
|  |  | 14:00 to 14:15 | 1.87 |
| Hydrogen Sulphide ($H_2S$) | 6 June 2000 | 10.00 to 10:10 | Nil |
|  |  | 11:00 to 11:20 | Nil |
|  |  | 11:30 to 1200 | 0.2 |
| Ammonia ($NH_3$) | 6 June 2000 | 10:20 to 10:30 | 15.45 |
|  |  | 12:15 to 12:25 | 25.75 |
|  |  | 13:00 to 13:15 | 17.13 |
| Chlorine ($Cl_2$) | 6 June 2000 | 10:40 to 10:55 | Nil |
|  |  | 12:40 to 12:55 | Nil |
|  |  | 13:30 to 13:45 | Nil |

Table 4.6: Analysis of air pollutants and data acquisition

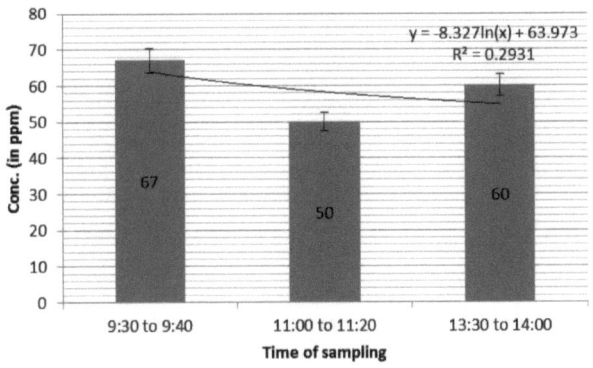

Figure 4.6.1: Carbon monoxide at Shadman Market

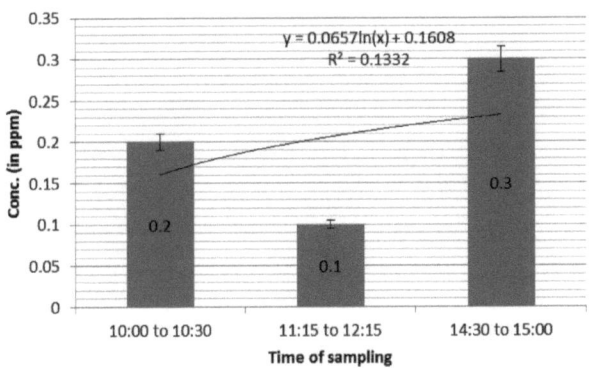

Figure 4.6.2: Sulphur dioxide at Shadman Market

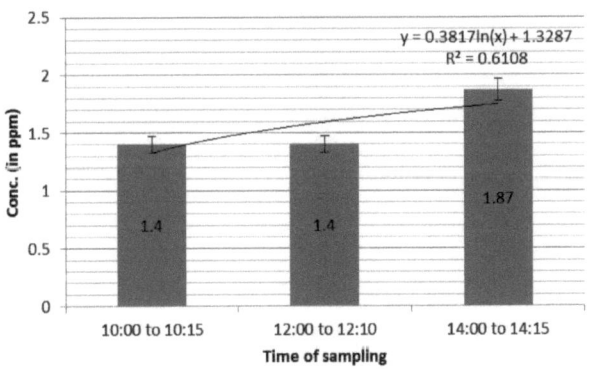

Figure 4.6.3: Nitrogen dioxide at Shadman Market

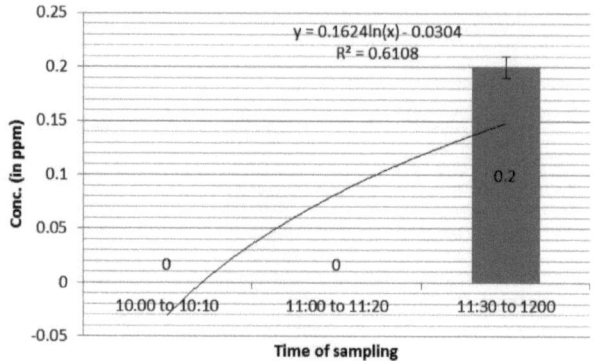

Figure 4.6.4: Hydrogen sulphide at Shadman Market

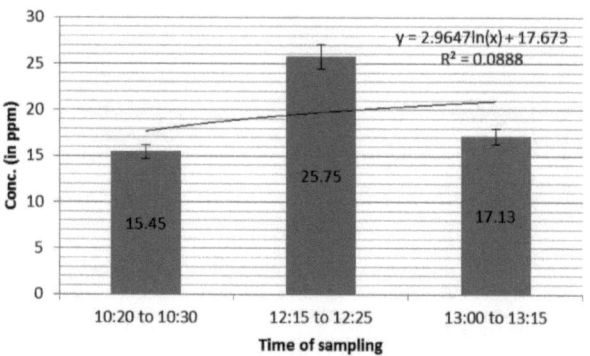

Figure 4.6.5: Ammonia at Shadman Market

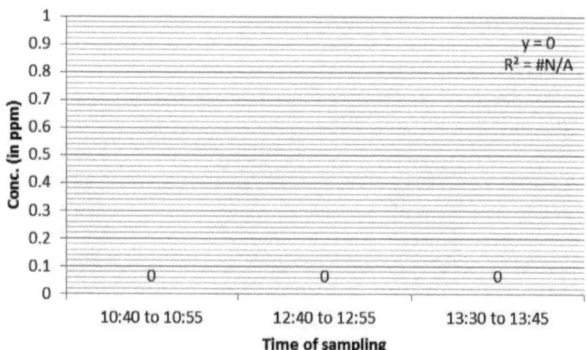

Figure 4.6.6: Chlorine at Shadman Market

### 4.10.3 Interpretation and trend analysis

The Shadman market is a commercial and residential area of Lahore. The concentration of CO, $SO_2$, $NO_2$, $H_2S$, $NH_3$ and $Cl_2$ has been measured at Shadman Market as shown in graphs.

As shown in graph 4.6.1, the concentration of carbon monoxide in the morning shows an increasing trend. At 9:30 in the morning, during office time and traffic density is higher. As the time passes the number of vehicles decreases, hence lowers the concentration. In afternoon the human activities and density of the vehicles increases which may give rise to the concentration level of carbon monoxide.

Sulphur dioxide also shows a sharp increasing trend. At 1:30pm as shown in the graph 4.6.2, its shows maximum concentration. It may be due to increase of vehicle density and with the sewage that is passing near the market.

The concentration of $NO_2$ in the morning shows an increasing trend. This may be due to the increase of temperature and vehicle with time producing nitrogen dioxide in large quantities. Due to the high pressure in the automobile engine, nitrogen and oxygen atoms in the air react and form this nitrogen oxide.

As shown in the graph 4.6.4, the concentration of hydrogen sulphide at Shadman market shows an increasing trend in the afternoon which may be due to the sewage passing near the market. This may be the main reason for the presence of hydrogen sulphide at this location.

The concentration of ammonia shows an increasing trend. It may be due to the sewage passing near the market and ammonia-containing compounds flowing in the sewage that produces ammonia in the large quantities at this location.

No colour has been developed in sampling chlorine. This may be due to two reasons; firstly the reagent formation could not be correct but the reagents were prepared a number of times, secondly, this may be due to the absence and presence in non-detectable range of chlorine at the location. After taking a number of observations the concentration of chlorine has not been detected.

## 4.11 Peak concentrations (in ppm)

| Sampling site | Carbon monoxide | Sulphur dioxide | Nitrogen dioxide | Hydrogen sulphide | Ammonia | Chlorine |
|---|---|---|---|---|---|---|
| Badami Bagh | 80 | 0.9 | 2.1 | 0.9 | 20.58 | 1.1 |
| Campus Bridge | 50 | 0.3 | 0.7 | No color | 15.45 | No color |
| Charing Cross | 67 | 1.3 | 1.4 | No color | 25.74 | No color |
| Ichhra Market | 75 | 1.3 | 0.93 | No color | 20.58 | No color |
| Main Market Gulberg | 50 | 0.3 | 0.7 | No color | 20.58 | No color |
| Railway Station | 75 | 2.6 | 2.45 | 0.5 | 25.74 | 0.9 |
| Shadman Chowk | 75 | 1.3 | 1.87 | No color | 20.58 | No color |
| Shadman Market | 67 | 0.3 | 1.87 | 0.2 | 25.74 | No color |
| Yadgar Chowk | 100 | 1.95 | 2.8 | 0.2 | 30.9 | 0.4 |
| Yateem Khana Chowk | 50 | 1.3 | 2.1 | No color | 17.13 | No color |

Table 4.7: Highest levels of gaseous pollutant concentration

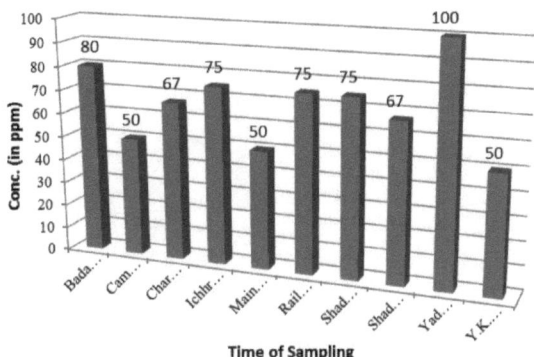

Figure 4.7.1: Maximun concentration of carbon monoxide

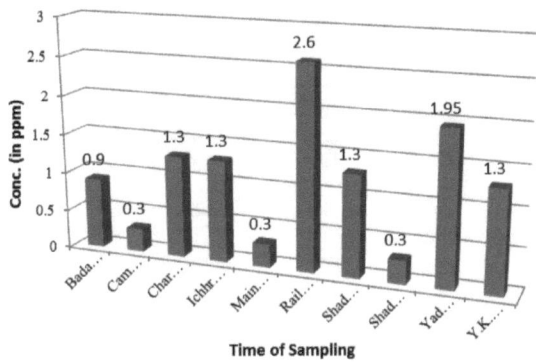

Figure 4.7.2: Maximun concentration of sulphur dioxide

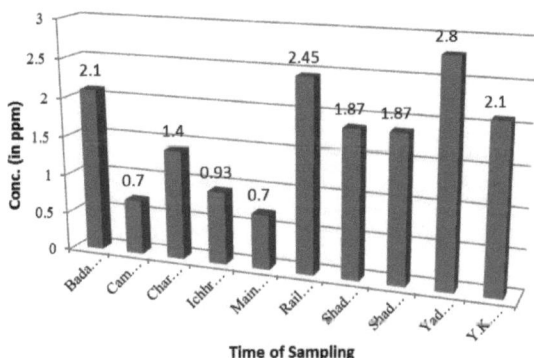

Figure 4.7.3: Maximun concentration of nitrogen dioxide

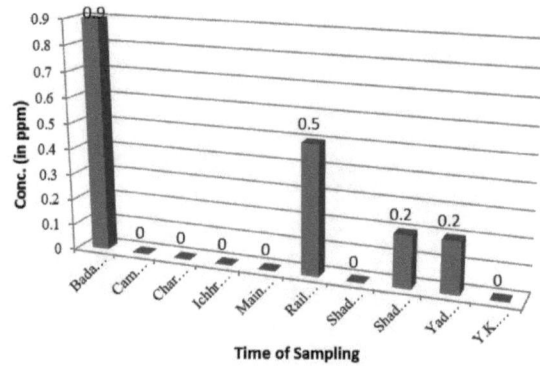

Figure 4.7.4: Maximun concentration of hydrogen sulphide

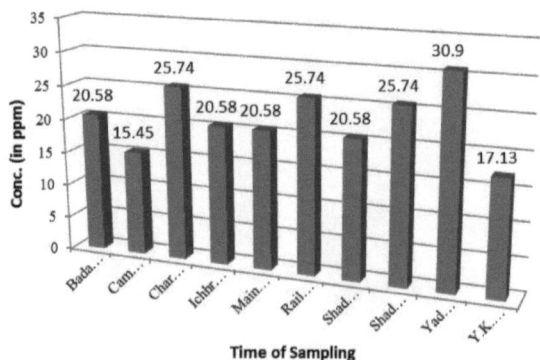

Figure 4.7.5: Maximun concentration of ammonia

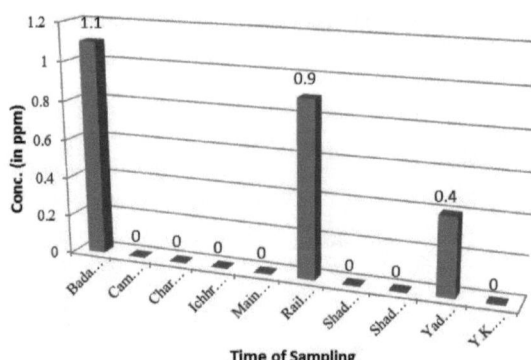

Figure 4.7.6: Maximun concentration of chlorine

### 4.11.1 Interpretation and trend analysis

The concentrations of CO, $SO_2$, $NO_2$, $H_2S$, $NH_3$ and $Cl_2$ have been measured at various commercial and residential locations of Lahore city as shown in the graphs where maximum observed concentrations have been plotted.

The concentration of carbon monoxide is found above the allowable thresholds at Badami Bagh, Ichhra Market, Railway Station, Shadman Chowk and Yadgar Chowk as shown in the graph. Whereas the areas of Campus Bridge, Main Market Gulberg and Yateem Khana Chowk has been found at allowable concentration of carbon monoxide.

The concentration of nitrogen dioxide is more than safe limits at Badami Bagh, Railway Station, Yadgar Chowk and Yateem Khana Chowk as shown in the graph. Whereas the areas of Campus Bridge, Main Market Gulberg and Ichhra Market has been found at allowable concentration of nitrogen dioxide.

Sulphur dioxide has been found at its maximum value at Shadman Chowk and Yadgar Chowk as shown in the graph. Whereas the areas of Campus Bridge, Main Market Gulberg and Shadman Market has been found at allowable concentration of sulphur dioxide.

Hydrogen sulphide has been found at its maximum value at Badami Bagh and Railway Station as shown in the graph. Whereas hydrogen sulphide has not be detected in the areas of Charing Cross, Campus Bridge, Main Market Gulberg and Ichhra Market.

Ammonia has been found at its maximum value at Charing Cross, Railway Station and Yadgar Chowk as shown in the graphs. Whereas the areas of Campus bridge and Yateem Khan Chowk are found at allowable concentration of ammonia in the air.

Chlorine has been found at its maximum value at Badami Bagh and Railway Station as show in the graph and is not found at Campus Bridge, Charing Cross, Main Market Gulberg, Shadman Chowk, Shadman Market and Yateem Khana Chowk. At Yadgar Chowk its concentration is in allowable limit.

## 4.12 Mean concentrations (in ppm)

| Sampling site | Carbon monoxide | Sulphur dioxide | Nitrogen dioxide | Hydrogen sulphide | Ammonia | Chlorine |
|---|---|---|---|---|---|---|
| Badami Bagh | 74.25 | 0.63 | 1.79 | 0.70 | 11.72 | 0.90 |
| Ichhra Market | 68.16 | 0.97 | 0.70 | Nil | 16.73 | Nil |
| Main Market Gulberg | 41.00 | 0.27 | 0.48 | Nil | 17.72 | Nil |
| Railway Station | 61.66 | 2.00 | 2.04 | 0.50 | 21.00 | 0.66 |
| Shadman Market | 59.00 | 0.20 | 1.55 | 0.20 | 19.44 | Nil |
| Campus Bridge | 36.00 | 0.25 | 0.63 | No colour | No colour | No colour |
| Charing Cross | 54.16 | 1.07 | 1.01 | No colour | 22.29 | No colour |
| Shadman Chowk | 64.00 | 0.97 | 1.40 | No colour | 17.15 | No colour |
| Yadgar Chowk | 78.00 | 1.31 | 2.25 | 0.20 | 25.32 | 0.30 |
| Yateem Khana Chowk | 41.00 | 0.97 | 1.79 | No colour | 15.44 | No colour |

Table 4.8: Mean values of levels of gaseous pollutant concentration

## 4.12.1 Mean of past pollutant concentrations (in ppm)

| Sampling Site | Carbon monoxide | Sulphur dioxide | Nitrogen dioxide | Hydrogen Sulphide | Ammonia | Chlorine |
|---|---|---|---|---|---|---|
| Campus Bridge | Not Monitored | 0.16 | 0.48 | Not Monitored | Not Monitored | Not Monitored |
| Charing Cross | 22.5 | 2.3 | Not Monitored | No colour | 18.3 | No colour |
| Railway Station | Not Monitored | 2.16 | Not Monitored | Not Monitored | Not Monitored | Not Monitored |
| Shadman Chowk | Not Monitored | 4.55 | No colour | Not Monitored | 25.18 | No colour |
| Yadgar Chowk | Not Monitored | 2.16 | 1.47 | Not Monitored | Not Monitored | Not Monitored |
| Yateem Khana Chowk | 22.5 | 4.8 | Not Monitored | Not Monitored | 15.12 | No colour |

Table 4.9: Mean values of levels of gaseous pollutant concentration of previous year

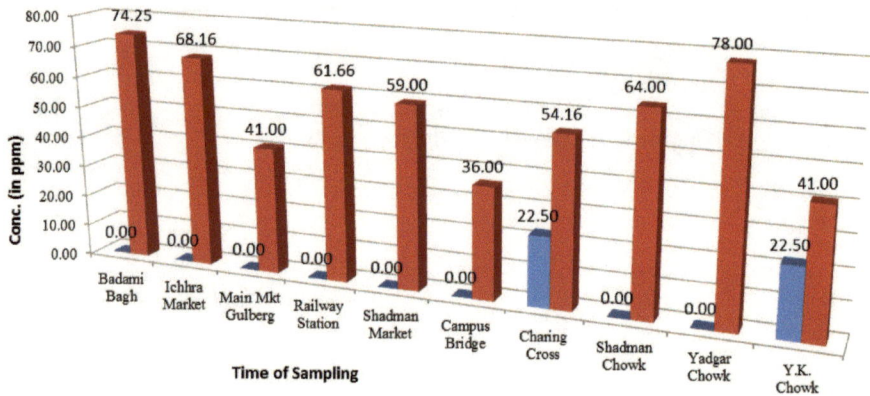

Figure 4.8.1: Comparison of mean concentrations of carbon monoxide. Blue concentrations are for year 1999 and red for 2000. A zero value indicate missing data.

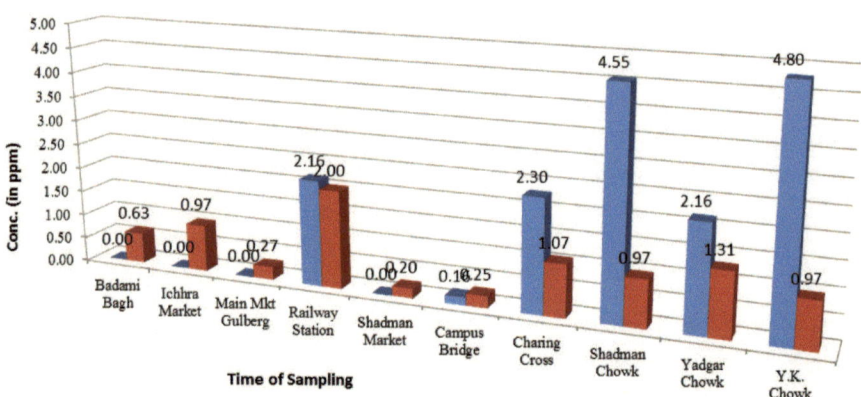

Figure 4.8.2: Comparison of mean concentrations of sulphur dioxide. Blue concentrations are for year 1999 and red for 2000. A zero value indicate missing data.

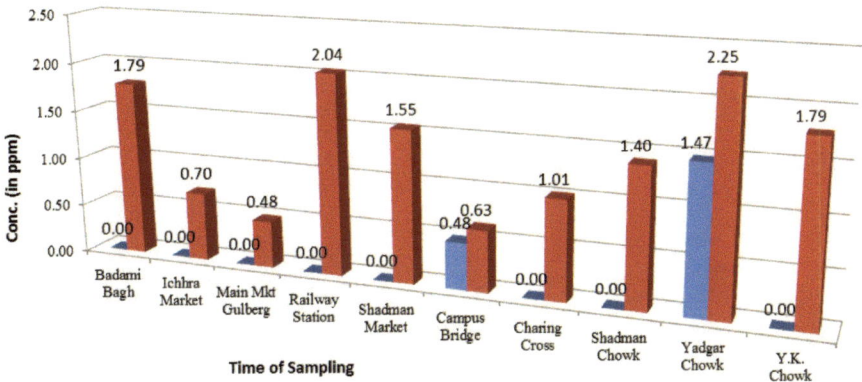

Figure 4.8.3: Comparison of mean concentrations of nitrogen dioxide. Blue concentrations are for year 1999 and red for 2000. A zero value indicate missing data.

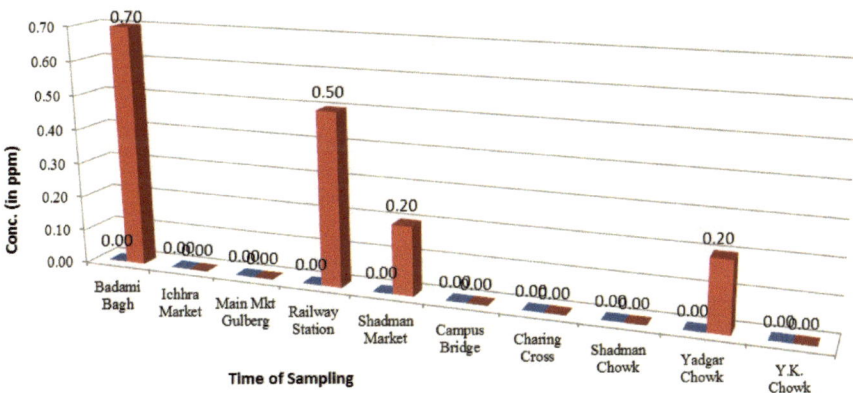

Figure 4.8.4: Comparison of mean concentrations of hydrogen sulphide. Blue concentrations are for year 1999 and red for 2000. A zero value indicate missing data.

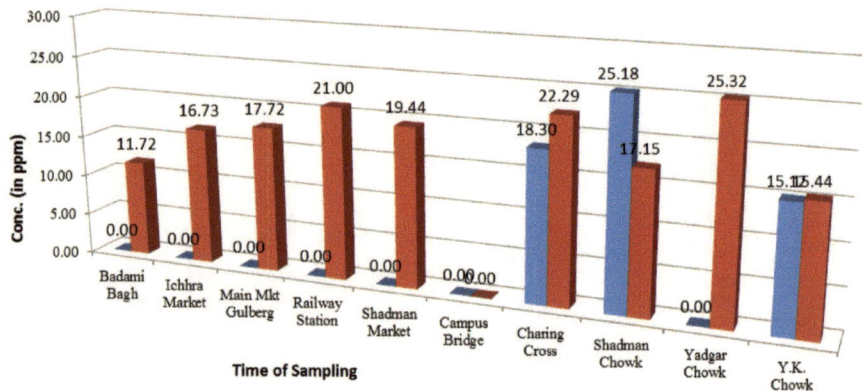

Figure 4.8.5: Comparison of mean concentrations of ammonia. Blue concentrations are for year 1999 and red for 2000. A zero value indicate missing data.

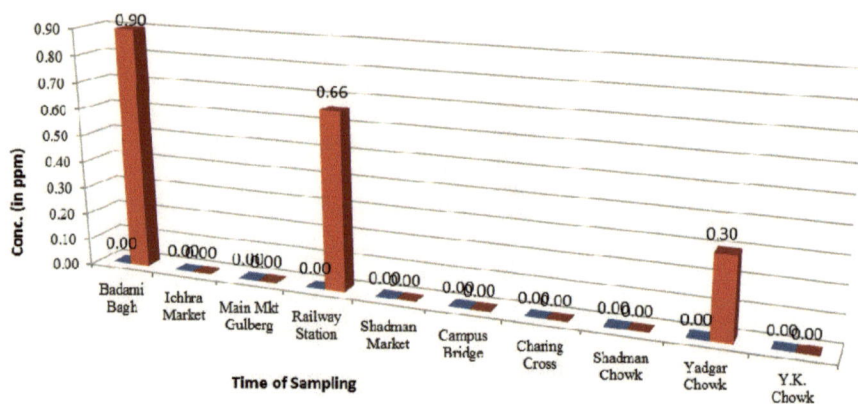

Figure 4.8.6: Comparison of mean concentrations of chlorine. Blue concentrations are for year 1999 and red for 2000. A zero value indicate missing data.

## 4.12.2 Interpretation and trend analysis

The concentrations of CO, $SO_2$, $NO_2$, $H_2S$, $NH_3$ and $Cl_2$ have been measured at various commercial and residential locations of Lahore city as shown in the graphs where mean observed concentrations have been plotted.

At Badami Bagh, as shown in the graph, all gases show an average value of concentration except carbon monoxide which is higher than the WHO standard, because it is highly polluted commercial and residential area and is affected by large number of light and heavy vehicles. Gases emitting from the garbage and a large amount of mud which has been formed as a result of rain over the soil waste present in the area also affect this area.

At Ichhra Market, the value of carbon monoxide is fairly higher than the specified range. This area is also affected by the passage of large number of low-grade diesel vehicles. Hydrogen sulphide and chlorine could not be detected at this location.

At Main Market Gulberg, the values of all the gases are within the specified range whereas the concentrations of some gases are not found like hydrogen sulphide and chlorine.

At Railway Station the concentration of carbon monoxide and ammonia are comparatively high, the mean concentration of carbon monoxide is 61.66 ppm whereas the mean concentration of ammonia is 21 ppm, but it is within the specified range.

At Shadman Market, the mean concentration of carbon monoxide is also higher than the allowable range; the value of ammonia is also higher but is in the allowable range. The concentration of hydrogen sulphide has been detected but chlorine has not been detected.

### 4.12.3 Interpretation and inter-comparison

The average of past data at various locations was found and is plotted in the graph against the present data of year 2000.

The concentrations of CO, $SO_2$, $NO_2$, $H_2S$, $NH_3$ and $Cl_2$ have been observed at various locations of Lahore and shown in the graph and have calculated the average of concentrations of these locations for Lahore data.

Carbon monoxide was not monitored last year at Campus Bridge, Railway Station, Shadman Chowk, and Yadgar Chowk. An increasing trend of CO has been observed at the Mall and Yateem Khana Chowk.

The concentration of sulphur dioxide was found to be decreasing at Shadman Chowk, The Mall, and Yateem Khana Chowk. This may be due to more smooth flow of traffic at the Mall and Shadman Chowk.

Nitrogen dioxide was not monitored last year at The Mall, Railway Station, Shadman Chowk and Yateem Khana Chowk therefore, a comparison could not been made. Whereas an increasing trend has been observed at Campus Bridge and Yadgar Chowk.

Hydrogen sulphide concentration was not monitored at any place last year. So a comparison with current year (2000) could not been made. In regard to this year data, it has been found at Railway Station and Yadgar Chowk. Whereas it has been attempted to sample hydrogen sulphide at The Mall, Campus Bridge, Shadman Chowk and Yateem Khana Chowk and it could not be measured in these areas.

The concentration level of ammonia shows increasing trend at The Mall and Yateem Khana Chowk and decreasing tend at Shadman Chowk. This may be due to more number of vehicles at Yateem Khana Chowk and smooth flow of traffic at Shadman Chowk.

The comparison relation of chlorine with the past data cannot be established because it was not measured at any of these places last year.

## 4.13 Lahore city averages (in ppm)

| Sampling site | Carbon monoxide | Sulphur dioxide | Nitrogen dioxide | Hydrogen sulphide | Ammonia | Chlorine |
|---|---|---|---|---|---|---|
| 1999 Average Data of Lahore City | 23.5 | 2.42 | 1.05 | Not Monitored | 18.97 | Not Monitored |
| 2000 Average Data of Lahore City | 57.72 | 0.86 | 1.464 | 2.98 | 18.53 | 0.62 |

Table 4.10: City averages values of levels of gaseous pollutant concentration

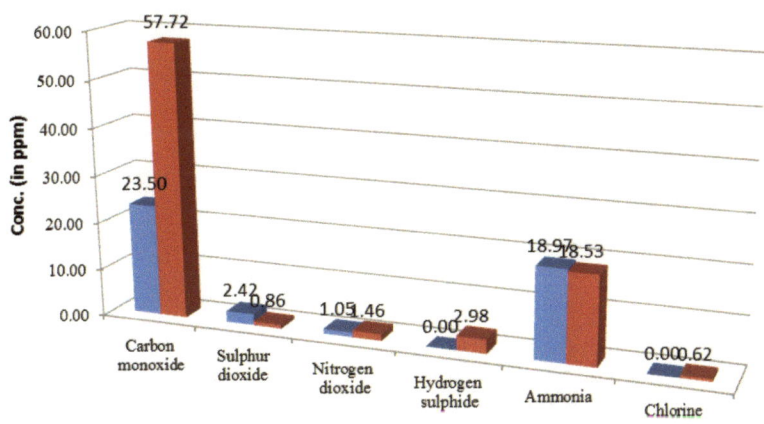

Figure 4.9: Overall comparison of average of all measurements in Lahore. Blue concentrations are for year 1999 and red for 2000. A zero value indicate missing data.

### 4.13.1   Interpretation and trend analysis

The average concentration of past data of Lahore are calculated and has been plotted in the graph against the average concentration of present data of Lahore.

The concentrations of CO, $SO_2$, $NO_2$, $H_2S$, $NH_3$ and $Cl_2$ have been observed at various locations of Lahore as shown in the graph and have calculate an overall average of concentration of air pollutants for Lahore.

The value of carbon monoxide shows an increasing trend because of increased number of activities in the city and large number of registered and non-registered vehicles.

The value of sulphur dioxide shows a decreasing trend. The value of nitrogen dioxide and ammonia is almost stable whereas the concentration level of hydrogen sulphide and chlorine was not monitored last year therefore, a comparison has not been made.

# Chapter 5:
# Air pollution control

## 5.1 Gaseous pollution control methods

Gaseous pollution control methods are classified into three types. They are as follows.

1). Absorption towers

2). Adsorption towers

3). Combustion

### 5.1.1 Absorption and adsorption towers

Absorption and adsorptions towers may be of two types; one is spray towers and other one is packed bed towers.

They are enclosed towers with inlet and outlet arrangements for both polluted gas and sorbing liquid (scrubbers). The polluted gas is sent to the tower through inlet. The gas is made to intimate contact with the spray of scrubbing liquid. The liquid absorbs one or more gas pollutants from polluted gas. The efficiency of this process depends on the following factors.

1). The amount of surface contact between gas and liquid.

2). Contact time

3). Concentration of the absorbing medium, and

4). Rate of reaction between the liquid and gases

### 5.1.2 Combustion

If exhaust gases are organic in nature, combustion process is like flame combustion (furnace) or catalytic combustion (flare) are used. Proper proportion of oxygen, temperature, time and turbulence are necessary for complete combustion.

Flame combustion is of two types. These are incinerators and after-burners. Incinerators are used for combustible gases, to develop a flame in the presence of proper amount of air. After-burners are used for complete combustion of the incinerator's effluent gas. It's operation cost is more and removes only combustible gases.

Catalytic combustion method is used for the combustion of gases whose combustibility is low flammable range and also the gases having low operating temperature e.g., effluent gases from refineries, paints etc. these require high initial cost and catalysts subjected to poisoning. Furnace combustion may be used to control methyl mercaptan, hydrogen sulphide and methyl sulphide odours from craft paper mill, vapour control from paints and varnishes.

## 5.2 Particulate pollution control methods

Methods of removal of particulate matter may be classified in five types. They are as follows.

1). Setting chambers
2). Cyclone separators
3). Wet collectors
4). Bag filters
5). Electrostatic precipitators

### 5.2.1 Settling chamber

These are closed tanks having inlets and outlet arrangements. The polluted gas is allowed to enter into the chambers, where the velocity of the gas is reduced sufficiently to minimize the turbulence, nearer to the laminar flow condition, which will allow the particulate to settle by gravity ad separates from the gas stream. To ensure uniform velocity, baffle or mesh screens may be suspended in the chamber. A hopper at the bottom with 1:1 slope is provided at the bottom of the settling zone to collect the settled particulate matter. The flow velocity is usually in the range of 0.5 – 3 m/s. the minimum size of the particle moved is 10 micron. Efficiency is high for the removal of particles greater than or equal to 50 microns. The advantages of using this method are low initial cost, low maintenance cost, simple structure, dry / continuous disposal of solid particles, low pressure drop and high efficiency for coarse particles. In this chamber, no moving parts are involved. The disadvantages of this method, on the other side, are requirement of more space, low efficiency for low concentration suspensions and lesser size of particles. Laminar conditions may not exist always.

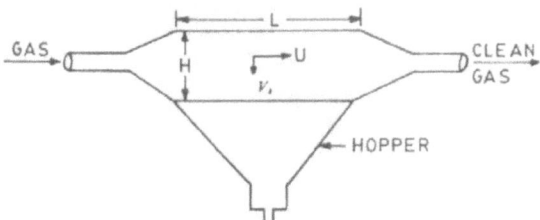

Figure 5.1: Settling chamber (Source: Wikipedia)

Here

Q = Flow rate of polluted gas

L = Length of the chamber

H = Depth of the chamber

U = Horizontal velocity component of the gas

$v$ = Settling velocity of particles

d' = Diameter of a particle

### 5.2.2 Cyclone separator

A cyclone separator is an enclosed vertically placed cylinder with an inverted cone at the base. The inlet is arranged with a tangential entry and an outlet arrangement at the top of cylinder.

The particulate containing gas enters the cylinder through the tangential inlet, which gives a whirl motion. The polluted gas pushes towards the periphery of the cylinder. Due to the centrifugal action, the heavy particles move towards the edge of the cyclone. There, the particles slide down into a conical collector (hopper bottom). The gas further proceeds and reaches the bottom of the conical section. There the gas moves upwards as a small inner spiral (due to cone base), which is concentric to the first spiral. After reaching the top of the cylinder, the clean gas leaves the cylinder through an outlet pipe.

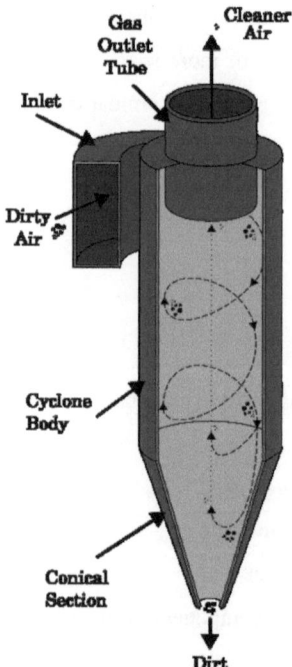

Figure 5.2: Cyclone separator (Source: Wikipedia)

The advantages of this method are low initial cost, low maintenance cost, simple in construction / design, no moving parts and less pressure drop (25 mm – 200 mm). It can be constructed with different types materials to meet high temperature and pressure. Dry and continuous disposal of solid particles is possible in the size range of 5 – 40 microns. The disadvantages are the equipment may be subjected to abrasion. It is has low efficiency for particles of sizes range 5 – 10 micron as well as low efficiency for low concentration suspensions.

### 5.2.3 Wet collectors (Scrubbers)

Wet collectors are the enclosed tanks with a separate inlet and outlet arrangement for polluted gas and a scrubbing liquid. The particulate are removed from the gas with the help of liquid (scrubber). Generally the water is used as the scrubbing liquid.

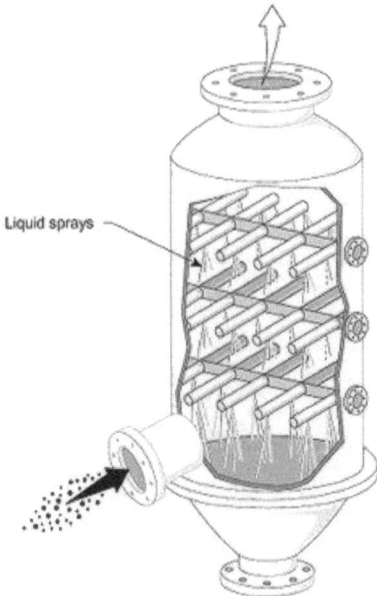

Figure 5.3: The working of a spray tower (Source: Wikipedia)

Separation of particle from the gas occurs when the particulates are made to strike a liquid surface within the wet collector. They can be used as precleaners. Depending upon the provisions of contact made between the gas and the liquid, the wet collectors are commonly divided four types:

In spray tower, generally the shape is either circular or rectangular and the particles of size range 1 – 2 microns are collected mainly by impingement. The efficiency is 94 – 99 % for the size range of 5 – 25 microns.

The cyclone scrubbers are similar to dry cyclone separators. Additionally it contains scrubbing spray systems with inlet and outlet arrangements. Water sprays are introduced in a variety of ways and the particles are collected by impingement and centrifugal action. It is effective for the removal of 1 -2 micron size particles. The efficiency is 90 – 98 % for the size range of 5 – 50 microns. The advantages of using cyclone scrubbers are low initial cost, high efficiencies for small particles and no particle retainment. It can remove the particulates and gas simultaneously. The disadvantages include waste disposal problems and high power consumption for high efficiency.

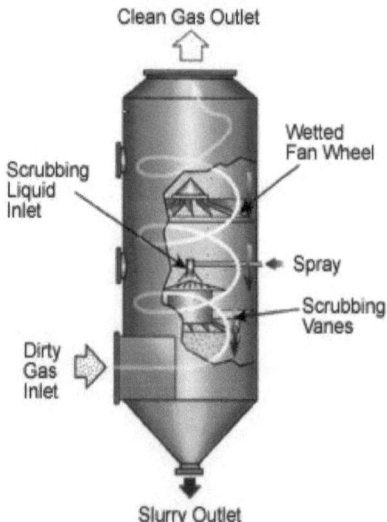

Figure 5.4: Cyclone scrubbers (Source: http://www.ustudy.in/node/3145)

Venturi scrubbers is a wet collector with a venturemeter shape. The gas is introduced through a venture tube at throat velocity of 60 to 100 m/s. the scrubbing liquid (water) sprays at just ahead of the Venturi throat. The particulars are collected along with the falling water. Its advantage is high efficiency for small particles. Where cleaning is difficult, this can be used due to compactness. The disadvantage is high pressure drop. Skilled personnel required for design and operation.

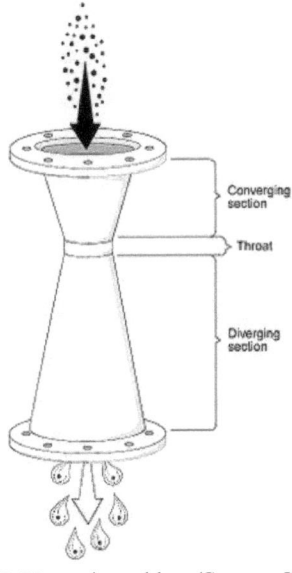

Figure 5.5: Venturi scrubber (Source: Wikipedia)

Packed bed tower is similar to spray towers with an additional facility for a pack bed, which is placed at height in a cylindrical tank. The inlet gas is allowed to pass through the bottom of the belt and while passing, an intimate contact between polluted gas and bed occurs. When the gas and particles reaches the top id the belt the spray of the liquid will catch the particle and carried away along with wash water. The size of particle are removed in the size of particles from 0.2 to 10 microns. The particulate and gases are removed simultaneously.

### 5.2.4 Bag filters

Bag filters are used for the removal of particles of size range of less than 10 micron. These are reliable, efficient system for particulate removal. These are arranged in an enclosure called as "bag house". The size of each bag is 120 to 200 meters in diameter and 2 to 10 meters long which are suspended. The outlet ends of the bags are open alternatively and attached to a manifold. The polluted gas enters through the inlet pipe. The larger particles will fall into hopper by gravity. The gas flow into the bags and leaves through the outlet pipe. The particulate matter retained on the inside of the bag and form a cake. The cake will be cleaned with the help of the shaking mechanism, causing the filter cake to be loosened. The loosened cake will fall into the hopper which is provided at the bottom of the bag house.

The advantages of using a bag house are high efficiency for small particles. The particles of even 0.01 micron can be effectively removed. Disposal of collected material is done in dry condition. The disadvantages are high pressure drops, high maintenance cost and filter operations upto moderate conditions. The clogging of filter takes place hence it requires frequent cleaning.

### 5.2.5 Electrostatic precipitators

Electrostatic precipitators are widely popular for the removal of very small size particulate matter. The polluted gas is allowed to pass between two electrodes one is negative charged high voltage electrode (e.g., wire) and the other is positively charged plate or a cylinder. High potential difference (25 to 100 kV) is maintained between them.

Because of high potential difference, a powerful ionizing field is formed. This creates an active glow zone (blue electric discharge) very close to negative electrode, which is called a Corona. As the negative ions migrate towards the collecting electrode (low potential electrode), they also charge the passing particulates. The electric field attracts the particulates towards the collecting electrode (plate) and deposited there. The advantages of electrostatic precipitators are high collecting efficiency (+99%), low pressure drop comparing with high efficiencies, handling of large volume of high temperature gas and collection of both wet and dry particles. It can remove very small particles even less than 0.01 microns and its treatment time is 1 – 10 seconds. Its disadvantages are relatively high initial cost, more power consumption and high maintenance cost. It is important that precautions be taken for the safety of the personnel from high voltage. Skilled personnel are required for the design, operation and maintenance. The ionization of gas occurs only in a limited operated range and fluctuations drastically reduce efficiency.

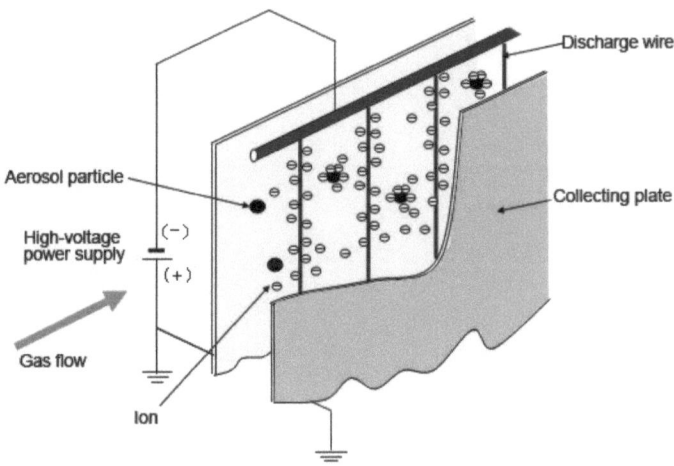

Figure 5.6: Electrostatic precipitator (Source: http://www.hitachi-infra.com.sg/)

# Chapter 6:
# Public survey, conclusions and recommendations

## 6.1 Public survey

A public survey has been conducted to know the opinion of the general public about air pollution. In this survey the questions asked are the most common and about the general knowledge of the air pollution.

A total of eighty people contributed in this survey and gave their response on the Performa specially provided containing the questions for which the answers are requested. The response is very encouraging. More than 80% people know the problem of air pollution and its effects to human health.

The eighty response forms have been received from all parts of the city, from people of different pollutant exposures and from people of different professions including physicians, surgeons, environmentalists, shopkeepers, drivers, conductors, traffic police constables and especially the people residing in the colonies nearer to the bus stops, industrial areas and high exposure areas. Thanks to all these persons for their time devotion and contribution towards compiling this survey.

Q. No. 1: Do you know about air pollution?

Answers/Responses

| Yes | 75 |
|-----|----|
| No  | 5  |

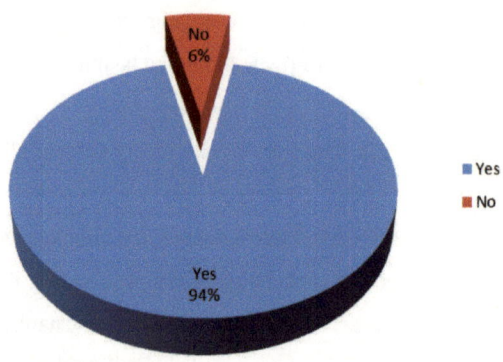

Q. No. 2: Which type of pollution is more dangerous?

Answers/Responses

| Air pollution | 45 |
|---|---|
| Water pollution | 10 |
| Noise pollution | 7 |
| Land contamination | 5 |

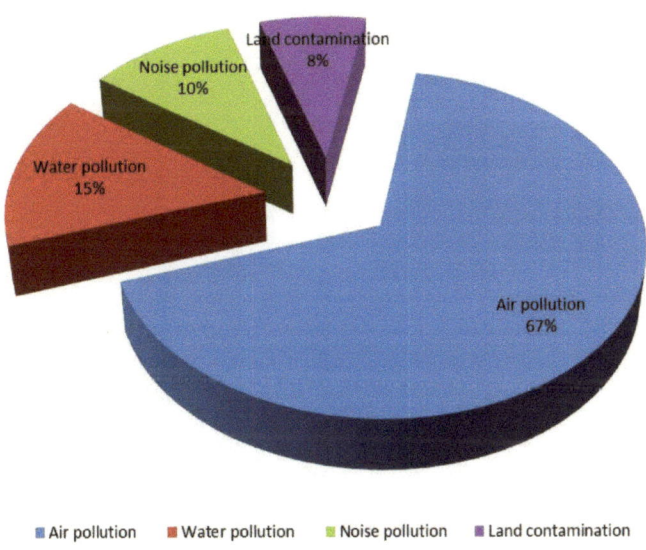

Q. No. 3: What do you think which gas is more harmful to human life?

Answers/Responses

| | |
|---|---|
| CO | 56 |
| $SO_2$ | 13 |
| $NO_x$ | 7 |

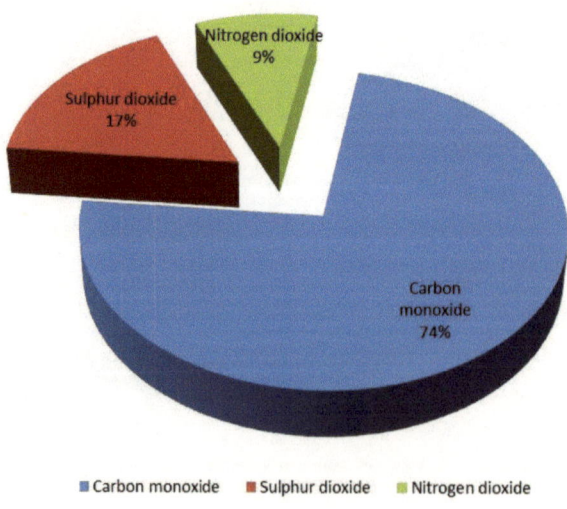

Q. No. 4: Which health effects are produced by the air pollution?

Answers/Responses

| Nose irritation | 12 |
| Eye irritation | 23 |
| Lung diseases | 24 |
| Asthma | 16 |
| Other | 2 |

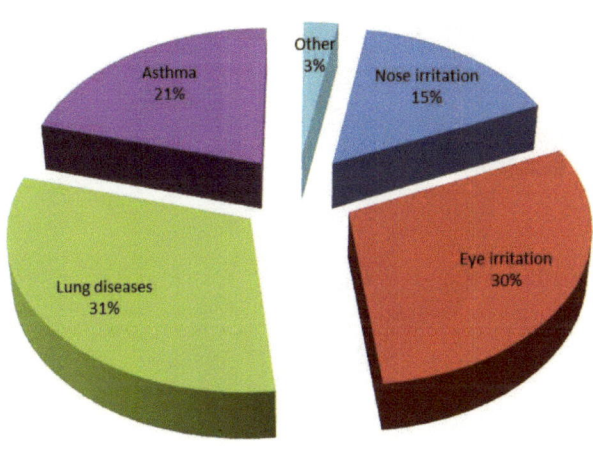

Q. No. 5: Do you think that air pollution has affected your property?

Answers/Responses

| Yes | 55 |
|-----|----|
| No  | 23 |

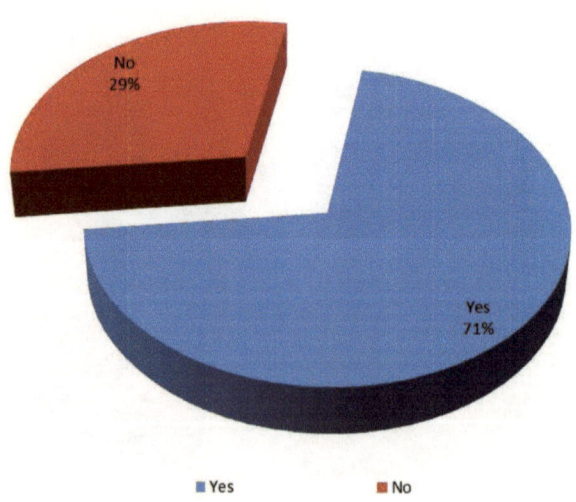

Q. No. 6: Do you think that air pollution has affected the atmosphere?

Answers/Responses

| Yes | 41 |
|-----|----|
| No  | 6  |

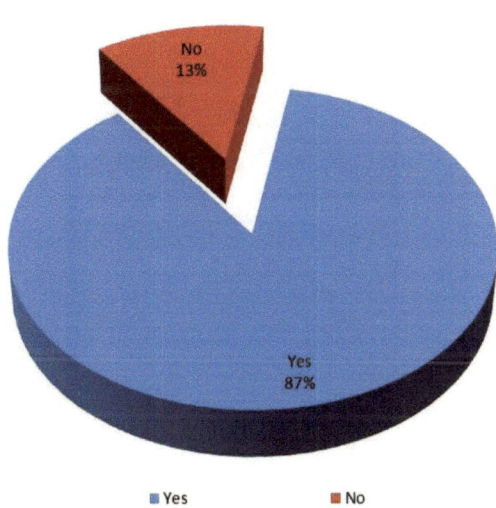

Q. No. 7: Does air pollution affects animals and plants?

Answers/Responses

| Yes | 58 |
|-----|----|
| No  | 6  |

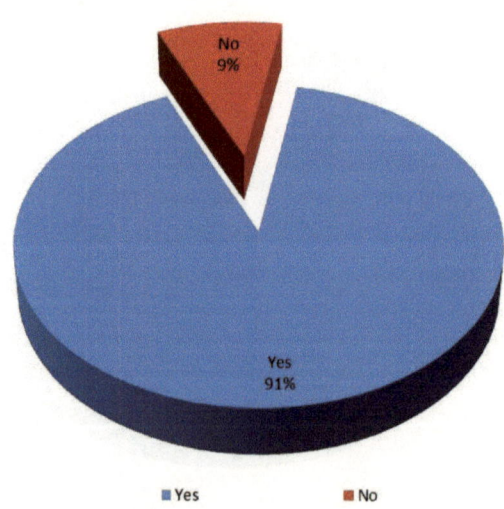

## 6.2 Conclusions and recommendations

1). To manage air quality in the future there should be an enhancement of coordination, at all levels of government, of transport, urban infrastructure and environmental management responsibilities. The coordination should include:

   a). the ongoing formal relationship embodied in the national environmental protection authorities and extensions such as the MoU between the authorities and the law enforcement commission
   
   b). integrated plans for optimal development of conurbation in terms of urban form and transit and transport systems including intermodal connections
   
   c). attention to pollution sources other than vehicles, especially among the agencies and levels of government operating within a particular air shed
   
   d). integrated planning systems which predict conditions conductive to high urban air pollution and provide that deliberate burns for hazard reduction are avoided in such periods

2). Air quality management measures, which receive the support of governments at commonwealth, state and municipal level, should be widely promoted through educational and information programs.

3). States, with the Commonwealth, should continue to collaborate in the ongoing development and implementation of nationally based methods for monitoring, data analysis, modelling and reporting. For this purpose they should continue to draw on the expertise resident within their own agencies like EPA's, meteorology departments and other agencies, universities and private consultancies.

4). Monitoring and air quality interpretation for the air-shed include the following:

- a). monitoring stations for the key pollutants located so as to sample the major zones of the air sheds. Their appropriateness of location should be determined by air shed modelling and data analysis
- b). data interpretation modelling and prediction moved to a GIS based methodology
- c). personal exposure estimates based on population demographics
- d). on a periodic basis studies of the origin species and size distribution of the particulate causing brown haze over major metropolitan air shed should be conducted
- e). the information obtained on airborne particulate and their sources should be used to modify or extend emission standards and air quality goals

5). Early adoption of inspection and maintenance programs based on vehicle testing, preferably at number of specialized centres located across the urban areas

- a). adoption when commercially available of onboard diagnostic systems
- b). support for programs for the acquisition of baseline information on the diesel fleet, its usage patterns and emission characteristics, including engine and fuel performance
- c). Development of in-service monitoring methods for diesel emissions, including particulate, should continue to be supported on a national basis
- d). co-operation by governments authorities in developing and implementing, as quickly as possible, an effective, low cost test to

identify those diesel powered vehicles which are excessive emitters of pollutants.

6). Governments and industry should jointly explore the potential emissions benefits that could flow from:

    a). reducing the sulphur content of diesel fuel to levels comparable with those planned in developed nations

    b). assessing the likely benefits of mandating the use of detergent additives in diesel fuel to combat the built-up detrimental deposits on fuel injectors

7). The use of CNG and LPG in all classes of vehicles should be encouraged through clear policies and use of gaseous fuels and by increase investment in refuelling infrastructures and vehicles in both the private and public sectors.

8). R&D and awareness programs should be maintained.

9). In relation to airports and aircrafts, adopt the following options
    a). minimize engine use while queuing and taxiing
    b). reduce auxiliary power unit use by encouraging airports to provide gates with power and preconditioned air.
    c). press for lower emissions from new engines as part of certification
    d). make air quality an essential part of the planning process for the siting of new airports

10). The applications of new technology intelligent transport systems (ITS) to the management and control of urban transport should pursued as a major priority in the country.

11). Demand management measures which enable individuals to make informed travel choices commensurate with their accessibility and mobility needs should be introduced. These include advanced traveller information systems, real time dynamic traffic control and information (e.g., on parking availability in areas). Traffic control measures aimed at improving freight and commercial vehicle operations should be adopted and vehicles using fuels with less potential for emissions should be encouraged and mandated

12). Reduction of Vehicle Kilometre Travelled (VKT) should be sought through urban planning and transit system design, which both allows and encourages the enhanced use of public transport and alternatives such as walking or bicycling for shorter journeys.

13). Local licences should be used to limit lead emissions level for the small number of affected sites. These could require covered storage and transfer points for concentrate storage and transfer.

14). Government could offer industry incentives for the location of new lead processing plants in regions away from urban areas.

15). Small to medium boiler and furnace operations should be encouraged to improve combustion performance and minimize pollutant emissions ($NO_X$, CO particulate) by means of

    a). incentive programs to reward organizations and maintaining combustion processes

    b). government supported testing services to help small boiler owners identity the need for particulate control especially for fine particles c). incentive schemes to promote the installation of fine particle collection equipment in small and intermediate plants

d). reinforcement of the greenhouse challenge initiative

16). Oil refineries in the air sheds to have licence performance conditions which meet urban requirements and demonstrate that adequate equipment is provided to cope with high sulphur crudes should the import of these become necessary.

17). New oil refineries and other potential $SO_x$ emitters should be sited outside urban air sheds.

18). Incentives to install more efficient sulphuric acid conversion plants should be provided.

19). Emissions of $NO_x$ and other pollutants from gas turbines in the urban air shed should be minimized. The actions should include

a). review of policies by governments that promote installation of gas turbines at locations that are unsatisfactory with respect to pollutant emissions

b). installation of larger more efficient gas turbines with reduced $NO_x$ emissions

c). incentives to minimize or preclude the use of liquid fuels in gas turbines located in urban areas

20). Specify vapour recovery requirements for environmental operating licenses for oil refineries and other affected industries such as car manufacturing printing spray painting and chemical manufacture.

21). For wood heaters all states and territories should adopt uniform legislation and coordinate their policies with respect to

    a). adoption forthwith of policies of emissions from domestic wood burning
    b). tightening of the current emission standard by 25% to 4 g/kg
    c). effective community education on correct heater operation
    d). restrictions on resale of old heater models which do not comply with standards
    e). an industry sponsored program to buy back old heaters
    f). local government targeting of very smoky heaters
    g). installation codes of practice
    h). controls to prevent the sale of excessively wet firewood
    i). media warning of high air pollution days so that people with alternative heating will not use open fireplaces or old model heaters
    j). banning of all new open fireplaces in urban areas

22). In any towns or cities where winter smoke is problem backyard burning should not be permitted

23). Hazard reduction burning should not be conducted in or near air sheds when weather patterns are conducive to smog formation

24). Alternative means of reducing forest litter and dry grass near urban areas should be sought

25). Encourage, with industry assistance, low $NO_x$ emission standards (of values equal to 25ppm or less) for all new water heaters and other domestic appliances.

26). Promote consumer awareness of the air quality benefits of actions such as

    a). use of low emission lawn mowers

    b). the use of water based acrylic paints including those with zero volatile matter

    c). minimizing household energy use

27). Adopt planning strategies which deliberately channel and concentrate additional population and non-polluting industry into specific zones or locations served by transport, employment and service facilities, either existing or new, so as to maximize self-containment of housing, jobs and services and minimize unproductive travel.

28). As well as improving the quality and choice of transit systems along the corridors, which serve such development, enhance the orbital connectivity between corridors and their major nodes.

29). Adopt a whole of life approach to building and dwelling design, construction and operation so as to optimize both the embodied energy incorporated in the building and its life time operating energy requirements.

30). Further pursue the implementation of energy efficiency in all new dwellings.

# Bibliography

Allem, A.C. (1997) *Roadside habitats: a missing link in the conservation.* The Environ., 17: 7-10.

Anderson, H. R. (1999) [Editors Holgate, S. T., H. S. Koren, et al. (1999)]. Health effects of air pollution episodes [*in "Air pollution and health"*], Academic Press. ISBN: 9780080526928. 1065pp.

Barker, J. R. and D. T. Tingey (1992). Air-pollution effects on biodiversity. Other Information: ManTech Environmental Technology, Inc., Corvallis, OR: 328pp.

Boer, B. (1996) *Plants as soil indicators along the Saudi coast of the Arabian Gulf.* Journal of Arid Environments, 33(1): 417-423.

Bolin, Bert, Göran Aspling, and Christer Persson. "*Residence time of atmospheric pollutants as dependent on source characteristics, atmospheric diffusion processes and sink mechanisms.*" Tellus 26.1-2 (1974): 185-195.

Bruce, N., Perez-Padilla, R., & Albalak, R. (2000). Indoor air pollution in developing countries: a major environmental and public health challenge. *Bulletin of the World Health Organization*, 78(9), 1078-1092.

http://en.wikipedia.org/wiki/Cyclonic_separation

Demographic Indicators (1998), Census Report, Pakistan Bureau of Statistics, Islamabad, Pakistan

Dockery, D. W., & Pope, C. A. (1994). Acute respiratory effects of particulate air pollution. *Annual review of public health*, 15(1), 107-132.

Dorney, J.R., G.R. Guntenspergen, J.R. Keough and F. Stearns. (1984) *"Composition and structure of an urban woody plant community"* Urban Ecol., 8(1): 69-90.

Esmen, N. A., & Corn, M. (1971). Residence time of particles in urban air. Atmospheric Environment (1967), 5(8), 571-578.

Fakhira Zahir (1997). *Spatial Analysis of Transport Network in Lahore City*. Department of Geography, University of the Punjab. MSc Thesis.

Field Instructions Manual (Undated), *Manual for air pollution monitoring methods*, Published by World Wildlife Fund (WWF)-Pakistan

Hussain, F. (1989) *"Field and Laboratory Manual of Plant Ecology"*. National Academy of Higher Education, University Grants Commission, Islamabad, Pakistan.

Isma Younes (2000) *"Spatial Patterns of Air Pollution in Lahore City"*, Department of Geography, University of the Punjab, Lahore, Pakistan (MSc Thesis).

Kazi, S. A., (1951), *Climatic Regions of West Pakistan*, Pakistan Geographical Review, 6(1), 1-22.

Khan, M. H., & Siddiqui, A. S. (1982). Growth and Fluctuations in the Output of Major Crops in Pakistan, 1950-51 to 1979-80. *The Pakistan Development Review*, 149-158.

Kupchella, C. E. and M. C. Hyland (1989). *Environmental Science: Living Within the System of Nature*, 2nd Edition. Allyn and Bacon. ISBN: 0205120164, 9780205120161. 637pp

Kupchella, C. E. and M. C. Hyland (1993). *Environmental Science: Living Within the System of Nature*, 3rd Edition. Prentice Hall. ISBN: 0132827409, 9780132827409. 579pp.

Miller G. Tyler (2000). *Living in the Environment: Principles, Connections and Solutions*. 11th Edition. Brooks / Cole. ISBN: 0534376088, 9780534376086. 815pp.

Nebel, B. J., & Wright, R. T. (1993). *Environmental science: the way the world works*. Prentice Hall Professional.

Maggs, R., Wahid, A., Shamsi, S. R. A., & Ashmore, M. R. (1995). Effects of ambient air pollution on wheat and rice yield in Pakistan. *Water, Air, and Soil Pollution*, 85(3), 1311-1316.

Mazhar, F., & Jamal, T. (2009). Temporal population growth of Lahore. *Journal of Scientific Research*, 39(1).

McClenahen, J.R. (1985) *"Community changes in a deciduous forest exposed to air pollution"*. Can. J. For. Res., 8(2): 432-438.

Mishra, A.K., K. Kamal, Jain and K. L. Garg. (1999) *"Role of higher plants in the deterioration of historic buildings"*. Sci. Total Environ., 167(3): 375-392.

Muller-Dombois, D. and H. Ellenberg. (1974). *"Aims and Methods of Vegetation Ecology"*. John Wiley & Sons. pp.55.

Samet, J.M., Marbury, Marian C., and Spengler, J.D. "*Health Effects and Sources of Indoor Air Pollution*" American Review of Respiratory Disease 1987; 136: 1486-1508.

Schwartz, J. (1994). Air pollution and daily mortality: a review and meta-analysis. *Environmental research*, 64(1), 36-52.

Smith, K. R. (1993). Fuel combustion, air pollution exposure, and health: the situation in developing countries. *Annual Review of Energy and the Environment*, 18(1), 529-566.

http://en.wikipedia.org/wiki/Spray_tower

Stern, A. C., Boubel, R. W., Turner, D. B., & Fox, D. L. (1984). *Fundamentals of air pollution*. 2$^{nd}$ Edition, Orlando. Academic Press. ISBN: 0-12-666580-X.

Streets, D. G., & Waldhoff, S. T. (2000). "Present and future emissions of air pollutants in China: SO2, NOx, and CO". *Atmospheric Environment*, 34(3), 363-374.

The Pakistan National Conservation Strategy (1999), Government of Pakistan

Thomas, W., Rühling, A., & Simon, H. (1984). Accumulation of airborne pollutants (PAH, chlorinated hydrocarbons, heavy metals) in various plant species and humus. *Environmental Pollution Series A, Ecological and Biological*, 36(4), 295-310.

Thurston, G. D., Ito, K., Kinney, P. L., & Lippmann, M. (1991). A multi-year study of air pollution and respiratory hospital admissions in three New York State metropolitan areas: results for 1988 and 1989 summers. *Journal of exposure analysis and environmental epidemiology*, 2(4), 429-450.

Wahid, A., Maggs, R. S. R. A., Shamsi, S. R. A., Bell, J. N. B., & Ashmore, M. R. (1995). Effects of air pollution on rice yield in the Pakistan Punjab. *Environmental Pollution*, 90(3), 323-329.

Wahid, A., Maggs, R. S. R. A., Shamsi, S. R. A., Bell, J. N. B., & Ashmore, M. R. (1995). Air pollution and its impacts on wheat yield in the Pakistan Punjab. *Environmental Pollution*, 88(2), 147-154.